Fundamental Theories of Physics

Volume 211

The international monograph series "Fundamental Theories of Physics" aims to stretch the boundaries of mainstream physics by clarifying and developing the theoretical and conceptual framework of physics and by applying it to a wide range of interdisciplinary scientific fields. Original contributions in well-established fields such as Quantum Physics, Relativity Theory, Cosmology, Quantum Field Theory, Statistical Mechanics and Nonlinear Dynamics are welcome. The series also provides a forum for non-conventional approaches to these fields. Publications should present new and promising ideas, with prospects for their further development, and carefully show how they connect to conventional views of the topic. Although the aim of this series is to go beyond established mainstream physics, a high profile and open-minded Editorial Board will evaluate all contributions carefully to ensure a high scientific standard.

Seiji Miyashita

Collapse of Metastability

Dynamics of First-Order Phase Transition

Springer

Seiji Miyashita
Department of Physics
University of Tokyo
Tokyo, Japan

ISSN 0168-1222 ISSN 2365-6425 (electronic)
Fundamental Theories of Physics
ISBN 978-981-19-6667-5 ISBN 978-981-19-6668-2 (eBook)
https://doi.org/10.1007/978-981-19-6668-2

This Springer imprint is published by the registered company Springer Nature Singapore Pte Ltd.
The registered company address is: 152 Beach Road, #21-01/04 Gateway East, Singapore 189721, Singapore

Preface

This book gives survey on the collapse of metastable state, i.e., dynamics of relaxation of metastable states by thermal and/or quantum fluctuation.

Phenomena in nature are governed by fundamental laws of physics. But, to understand what occurs in nature, cooperative behavior of many elements is essential. For example, even with the same elements, there are several phases, such as vapor, water, and ice of water molecules. To understand phenomena in nature, not only static properties in equilibrium, but also dynamical properties in processes under changing parameters play important roles.

As P. W. Anderson pointed out in the famous article "More is different", interactions among elements make ordered structures and symmetry breaking among them causes important phenomena which bring new intrinsic fundamental developments of view of physics. Although true symmetry breaking occurs in the thermodynamic limit associated with a phase transition, phenomena with apparent symmetry breaking occur in various time scales.

The variety of nature is owing to the metastability of such symmetry breaking phenomena with finite surviving time. In particular, hysteresis phenomena diversify properties of the nature. In hysteresis phenomena, a state persists as a metastable state even after some other state becomes more stable. To understand the mechanism of metastable state is one of key ingredients to understand the variety in natural phenomena.

The metastability collapses when the local stability of the metastable state disappears, e.g., the external field reaches to the so-called spinodal point. Although the spinodal point is defined as a critical point only in the special systems, i.e., the long-range interaction model in the thermodynamic limit, the spinodal like phenomena also occur for all metastable states with finite relaxation time, as a point where the relaxation time changes largely.

Mechanisms of relaxations from metastable state have been studied as a stochastic process jumping over the energy barrier by thermal fluctuation in classical systems as originally studied by Kramers and also by the work of Langer, and also as a quantum tunneling process through energy barrier studied by the WKB approximation and also methods using the idea of instanton for the field theorem (Coleman). In this book, we survey such phenomena and methods.

Metastability of thermodynamic phases appears associated with the first-order phase transition. We first study typical treatments of phase transitions by the mean-field theory where we study the free energy landscape which gives intuitive picture of barrier between states and also the spinodal phenomena. Then, we study how to treat the escape from the metastable states focusing the Kramers' method.

Because, at the first-order phase transition point, the system changes from one phase to the other, the configuration exhibits large changes in space. We give a brief review on spatial development of the ordering where the dynamical process depends on whether the density of order parameter is conserved or not.

We study nature of metastability from a viewpoint of eigenvalue problem of time evolution of the system given by the master equation. We demonstrate dependence of the relaxation time given by the eigenvalue as functions of the temperature and field for the kinetic Ising model. The concept of metastable state given by Langer is also reviewed.

The metastability in quantum system is also studied in the structure of eigenenergies of the Hamiltonian. First, we study a simple model with a double-well potential and find a peculiar structure related to the metastability, which we call metastable branch. The metastable branch is also studied in the model of particle conveyance, in which we carry a particle by moving a trap potential. This type of problems of particle escape by quantum tunneling has been studied by WKB method and also more sophisticated methods, e.g., those reviewed in the textbook of Child, which are also briefly reviewed. As a typical example, we introduce the quantum Stoner-Wohlfarth model.

For the transition to the stable state, some dissipative mechanism is necessary besides the tunneling. We study the quantum master equation and study dynamics of order parameter in dissipative environments, where Caldeira–Leggett mechanism is studied in a magnetic system. To understand the difference between quantum and thermal fluctuation, we give space–time configurations of quantum path-integral representation given by a quantum Monte Carlo simulation. We also study the first-order phase transition and related metastable states in the dynamical process of the optical bistability.

Finally, we review an attempt to quantitative estimations of relaxation time, i.e., the coercive force of real permanent magnets.

Related basic methods are given in Appendices for the convenience of readers.

I sincere thank Prof. Masuo Suzuki for his very kind guidance to the statistical physics. I appreciate Keiji Saito, Takashi Mori, Satoshi Hirosawa, Tatsuhiko Shirai, Satoshi Morita, Yoshiaki Teranishi, Hiroshi Jinnai, Takeo Ohta, Masamichi Nishino, Yuta Toga, Gen Tatara, Naoto Nagaosa and Bernard Barbara for their valuable comments and kind suggestions, and also Mr. Akiyuki Tokuno for his invitation and kind supports to write the book. Finally, I also thank my wife Ami for her continuous supports.

Tokyo, Japan Seiji Miyashita
June 2022

Contents

Chapter 1
Introduction

1.1 Concept of Metastability

The intuitive picture of metastable state is depicted as shown in Fig. 1.1. The ordinate is the energy when we discuss metastability of microscopic systems. However, at finite temperatures, the entropy effect must be included to study thermal stability, the ordinate should be the free energy. The abscissa M is a quantity which specifies the state, such as the magnetization of magnets.

When two states have the same energy, they coexist as depicted in Fig. 1.1a, When a bias field is applied, one of them has lower energy and it becomes the stable state, and the other becomes metastable as depicted in Fig. 1.1b. The metastable state is locally stable. Between the metastable state and the stable state, an energy barrier exists.

When the bias field increases, the energy barrier decreases and finally disappears at the spinodal point which is depicted in Fig. 1.1c. At this point, the metastable state becomes unstable and metastability disappears, and the state suddenly changes to the stable state, which causes a huge disturbance of configuration of state. This is the spinodal point. Above this point, the potential has single minimum, and no metastable state exists.

When the bias field is sweep back to the opposite direction, the state in the left valley persists until the other spinodal point with opposite sign. Thus, the state depends on history of the bias field. By a loop change of bias over the metastable region causes a loop of M, which is called hysteresis.

It should be noted that, even when the energy barrier exists, the metastable state can escape to the stable state by jumping over the barrier by thermal fluctuation or by quantum tunneling through the barrier. The escape rate is not zero as long as the barrier is finite.

S. Miyashita, *Collapse of Metastability*, Fundamental Theories of Physics 211,
https://doi.org/10.1007/978-981-19-6668-2_1

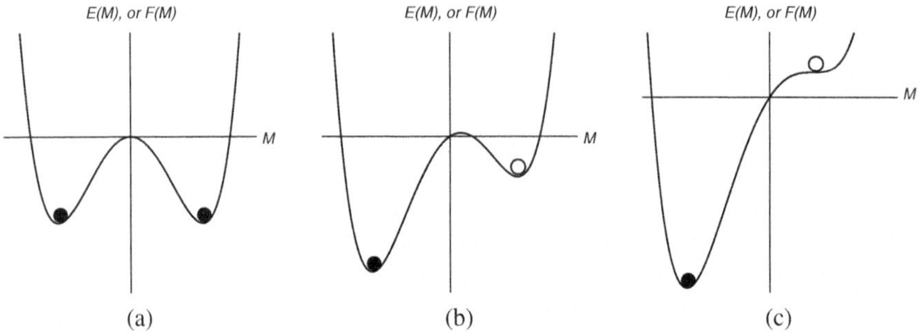

Fig. 1.1 Potential picture of **a** the coexistence, **b** metastability, and **c** the spinodal point. The closed circle denotes the stable state, and the open circle denotes the metastable state. The axis of abscissas is the order parameter of the system M, e.g., the magnetization for ferromagnetic systems

1.1.1 Life Time of a Metastable State

In finite-size systems, the energy barrier is finite, a metastable state relaxes to the stable state by some disturbance, and thus, it has a finite lifetime although it is locally stable. Thermal fluctuation or/and quantum fluctuation take the role of disturbance. Even if the size of system is infinite, when the interaction range is finite, the breakdown of metastable state can occur locally, for which the energy barrier is finite. Thus, the metastable state relaxes in a finite time. This process is called nucleation, which will be studied in Sect. 3.5.

In the present book, we will study how the metastable state relaxes to the stable state in various cases. In Chap. 2, we study metastability due to the first-order phase transition. The phase transition is one of the key ingredients of variety of nature and is an important topic of statistical physics. For the second-order phase transition and the associated critical phenomena, extensive studies have been done, and various important concepts such as universality class have been clarified. At the second-order phase transition, fluctuation of the order parameter diverges, and the so-called critical slowing down occurs. But the metastable state is not associated.

Metastable states appear being associated with the first-order phase transition. At the transition point of the first-order phase transition, two macroscopically different phases coexist, and the free energies of them are degenerate. For example, at the boiling point, the water and vapor coexist. If the parameters of the system, e.g., the temperature and pressure, cross the transition point, the equilibrium state, i.e., thermodynamically stable state, changes. When the system crosses the transition point, macroscopic jumps of quantities occur, such as the latent heat. But the metastable phase is still locally stable and persists even when the parameter exceeds the first-order phase transition point. Thus, practically no change of state is found at the critical point. Thus, it is difficult to locate the first-order phase transition point in a practical observation with changing the parameters.

The nature of the first-order phase transition is not well understood compared to the second-order transition because of the difficulty to treat the discontinuous change at the phase transition point. The understanding for such properties is still primitive, and it has been rather phenomenological. Indeed, the metastable state is not true stable state and thus it should relax to the equilibrium state by some way. However, we do not know any definite formula for the process in contrast to the static thermodynamic properties which can be described by the canonical ensemble method. Nevertheless, the relaxation of the metastable state plays important roles in various physical phenomena. Thus, we dare to challenge this problem.

In Chap. 2, we study the ferromagnetic Ising model by a mean-field analysis and discuss behavior of the free energy as a function of the order parameter. Metastable state appears below the critical temperature, and there the spontaneous magnetization appears and the two ordered states with opposite magnetizations coexist at zero magnetic field. In the presence of the field, one of them becomes stable and the other becomes metastable. In the mean-field analysis, this metastable situation is well described by the free energy, where the metastable state is given by a local minimum and the end point of metastable state gives the spinodal point.

In the Ising model the order parameter, i.e., the magnetization, is a scalar variable. But, in general, the magnetization can change its direction. In such cases, e.g., the uniaxial Heisenberg model, the metastable magnetization may change to the stable state by rotation. As a simplified model for magnetization reversal by rotation, the Stoner–Wohlfarth model is explained. Besides the first-order phase transition by the change of magnetic field, models which exhibit first-order phase transitions as a function of temperatures are also introduced. The gas–liquid phase transition is the most familiar phase transition in daily life, and the transition is also explained.

The mean-field analysis denotes the property in thermodynamic limit of uniformly ordered state, and the free energy barrier is macroscopic. Therefore, the lifetime of metastable state is infinite. In finite-size systems the barrier is finite, and the lifetime of metastable state is finite. The decay of metastable state is studied as a function of size in Chap. 3.

The decay is induced by escape of the population of metastable state to the stable state by jumping over the free energy barrier. The process is due to the thermal fluctuation and the rate is given by the Arrhenius law. The pioneering work of Kramers [1] is reviewed and it is applied to the mean-field model of ferromagnetic Ising model.

Even if the size of the system is infinite, in short-range interacting systems, the so-called nucleation, for which the energy barrier is finite, breaks the order locally. In such cases, the lifetime of metastable state is always finite, and thus the metastable state is not well-defined. Therefore, the spinodal point cannot be defined. However, metastable states and hysteresis are common phenomena in nature, and an idea of 'the dynamical spinodal point' is explained as a practical spinodal point in Sect. 3.6.

Because the state changes very largely at first-order phase transition, there appear characteristic spatial patterns of configuration in the process of the change. The Ising model describes both a uniaxial magnetic system and the gas-liquid transition and thermodynamic properties of the phase transition are the same in both systems. However, the ordering processes in the systems are very different. In the former, each local quantity, i.e., spin, can flip independently and thus the total magnetization is not conserved. In contrast, in the later, density of atoms cannot change independently, and it follows the equation of continuity to conserve the total number of atoms. Characteristics of dynamics of configurations during the phase change are reviewed in Chap. 4.

In Chap. 5, we survey the phase transitions from a view point of an eigenvalue problem. For static properties, the eigenvalue of the transfer matrix is discussed. The coexistence states are expressed by a quasi-degeneracy of eigenstates. The eigenvalues of transfer matrix give the free energy and the correlation length of order parameter. We characterize the singularity of free energy at first-order phase transition in the structure of eigenvalues as a function of parameters of the Hamiltonian, e.g., the temperature and the magnetic field. Dynamical properties are analyzed by studying the eigenvalues of time-evolution operator of the master equation.

Characterization of the singularity of free energy at first-order phase transition was studied by Langer [2]. He pointed out that the singularity appears as a function of the external field which is conjugate to the order parameter and the singularity is "so weak as to be experimentally unobservable". He pointed out that contribution from a large cluster of nucleation to the free energy has a singular dependence on the field, which can be regarded as an imaginary component in an analytic continuation. He also demonstrated that such singularity appears in the mean-field type partition function. Langer also related the singularity to the relaxation time of the metastable state [3]. These Langer's pictures give the essence of the singularity at the first-order phase transition, which are reviewed in Sect. 5.6. The picture has been argued in many succeeding studies [4–7].

Metastability in quantum systems is also an important issue, which is studied in Chap. 6. First, we study a system of double-well potential (Sect. 6.2). The system is characterized by the frequency ω of oscillations in the bottom of potential, the energy barrier ΔE, and the strength of quantum fluctuation Q. The separation of eigenenergies in each valley of potential is given by $\hbar\omega$. The states in both valleys hybridize and form nearly degenerate states with a small gap which depends exponentially on the energy barrier. Thus, depending on the ratios among these parameters, various cases appear. If $\hbar\omega$ and ΔE are of the same order and bigger than the gap, there is a few bound states in the well. On the other hand, if ΔE much larger than $\hbar\omega$, there are many bound states in the well, which corresponds to a situation of classical picture of metastability. We characterize the metastability from the structure of eigenvalues of the Hamiltonian. The metastable state is expressed as a 'metastable branch' which appears in energy level diagram as a function of a parameter of the Hamiltonian (Sect. 6.3). We analyze the eigenvalues in the problem of particle conveyance by a trap potential in Sect. 6.4. In Sect. 6.5, we study metastability of quantum magnets, e.g. single-molecular magnets in which quantum tunneling gives an interesting step-like structure of magnetization curve as a function of sweeping field. There, the effect of dissipation is also discussed. The spinodal points in quantum systems is studied as a problem of quantum Stoner-Wohlfarth model in Sects. 6.10 and 6.11. Comparison of quantum and thermal fluctuation is visualized in the transverse Ising model by making use of configurations obtained by a quantum Monte Carlo method in Sect. 6.12.

Besides the phase transitions in equilibrium states, there are dynamical first-order phase transitions in stational non-equilibrium systems. As an example, a cavity system driven by an external AC field (cavity QED) is studied in Sect. 6.13. The transmission coefficient of a cavity system changes discontinuously as a function of strength of input beam, which is called optical bistability (Sect. 6.14). Its mechanism is studied by a quantum master equation. The system shows a hysteresis under a sweeping field, and it shows curious structures of limit cycle in a periodically modulated AC external field (Sect. 6.15).

Finally, in Chap. 7, we present a trial for quantitative evaluation of relaxation time in real permanent magnets. There, persistence of metastable state is called 'coercivity'. For the coercivity at finite temperature, dynamics of an atomistic model is studied by the Stochastic LLG equation and also by a Monte Carlo method. The effect of the dipole-dipole interaction is also studied. Moreover, coercivity of a system consisting of ensemble of small crystals is also studied.

In the present book, we use several theoretical tools, e.g., mean-field approximation, mater equation, WKB approximation, quantum master equation, and so on. Ideas of the tools are briefly explained in Appendices.

Part I
Metastability in Classical Systems

Chapter 2
Metastability in Thermodynamic Systems

2.1 Introduction

Metastable state appears associated with the first-order phase transition. The macroscopic feature of metastability is captured by the idea of free energy landscape as a function of the order parameter. This picture is based on the mean-field theorem, in which the thermodynamic properties are expressed by a macroscopic order parameter. There, the spinodal point appears as an end of metastability. Therefore, we can clearly define the region of metastability.

We first briefly review mean-field analysis of the first-order phase transition. The most familiar first-order phase transition is the gas–liquid transition, for which we easily have an intuitive picture of transition, e.g., the boiling phenomena. But, here, for the simplicity of description, we study a ferromagnetic Ising model in Sect. 2.2. For the model, the order parameter is the uniform magnetization m. The model shows a first-order phase transition as a function of the magnetic field H below the critical temperature ($T < T_C$). There, we obtain hysteresis of magnetization, which is a consequent of the metastability, and analyze it by using a free energy $f(m; T, H)$ as a function of order parameter m at given temperature T and field H. At the end of metastability, i.e., at the spinodal point, the magnetization suddenly changes the sign and jumps to the stable state, and a discontinuous transition occurs from a metastable state to the stable state.

In the mean-field treatment, lifetime of metastable states is infinite because the energy barrier is macroscopic, i.e., proportional to the system size. Thus, the spinodal point is given as a well-defined singular point.

The order parameter, i.e., the magnetization m is a scalar variable, and when it changes the sign during the transition, the value of m must vanish. On the other hand, in magnets with three components $\boldsymbol{m} = (m_x, m_y, m_z)$, e.g., the Heisenberg model with uniaxial anisotropy, the magnetization $\boldsymbol{m}$ may change the direction by a rotation. In this case, the magnetization $\boldsymbol{m}$ shows a sudden change of its angle at a threshold of the field. A single magnet with uniaxial anisotropy is called Stoner–Wohlfarth model and is a basic model of magnetic systems, which is explained in Sect. 2.3.

S. Miyashita, *Collapse of Metastability*, Fundamental Theories of Physics 211,
https://doi.org/10.1007/978-981-19-6668-2_2

The above models exhibit first-order phase transitions as a function of the external field. We also introduce some examples which exhibit the first-order phase transition as a function of temperature in Sect. 2.4.

General idea of analysis of phase transition using of the free energy $f(m; T, H)$ as a function of order parameter is called Landau theory and a polynomial form of $f(m; T, H)$ is called 'Landau phenomenological free energy' which is explained in Sect. 2.5.

It is known that phase transitions in the ferromagnetic Ising model and in the gas-liquid system are the same type which is explained in Sect. 2.6.

In the present chapter, we study the system by a mean-field analysis, which corresponds to systems with long-range interaction and the size of system is infinite. Collapse of magnetization in the metastable state occurs by the thermal fluctuation in finite size systems. Even in infinite size systems, as in usual magnetic systems, when interactions among spins are short-range, the metastable state is destroyed by a local break-down of the order, i.e., the nucleation process. In this case, the singularity of spinodal point disappears. Nevertheless, a kind of spinodal phenomena appears. This problem is explained in the next Chap. 3.

2.2 Mean-Field Theory for a Ferromagnetic Ising System

In this section, we briefly review phenomenological approaches to the metastability adopting a ferromagnetic Ising model [8] as a typical model showing a phase transition. The Hamiltonian of the model in a given lattice, e.g., the square lattice (Fig. 2.1), is

$$\mathcal{H} = -J \sum_{<ij>} \sigma_i \sigma_j - H \sum_{i=1}^{N} \sigma_i, \quad \sigma = \pm 1, \tag{2.1}$$

where the first term denotes the exchange coupling between the nearest neighbor sites on the lattice and J denotes the strength of the coupling. The second term is the interaction with the magnetic field H (Zeeman energy).

This model was originally introduced to study uniaxial magnets, but it can represent all the systems with two-state variable, e.g., binary alloys where σ_i denotes the type of atoms, adsorption phenomena where σ_i denotes whether adsorbed atom exists or not on a site, the lattice gas model for the gas-liquid system where σ_i denotes whether an atom exists or not in a cell of space, spin-crossover system where σ_i denotes the high-spin state or the low-spin state, and opinion distribution (yeh or nay) in social problems. And thus, the model represents a rather universal mechanism.

Usually, properties of the first-order phase transition are discussed by using a phenomenological free energy $f(m; T, H)$ as a function of the order parameter m. This type of approach is called Landau theory (for general arguments, see Sect. 2.5), where the free energy is given in a polynomial form:

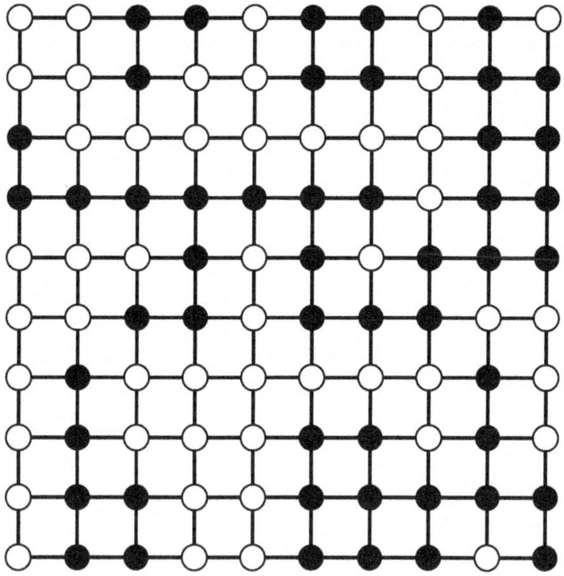

Fig. 2.1 Ising model on the square lattice. An example of configuration is depicted. Open circles denote $\sigma_i = 1$ and closed circles $\sigma_i = -1$

$$f(m; T, H) = am^2 + bm^4 + \cdots - Hm, \tag{2.2}$$

where T and H are the temperature and the magnetic field, respectively, and the coefficients are functions of them. The phenomenological free energy $f(m; T, H)$ corresponds to the free energy obtained in the mean-field analysis. Although most of the realistic models have short-range interactions, in the mean-field approximation, the system is regarded to have an effective long-range interaction. Nevertheless, a rough picture of the phase transition of the models is intuitively given by the mean-field analysis. It should be noted that the free energy $f(T, H)$ in the thermodynamics is a function of parameters of the Hamiltonian, i.e., T and H, while $f(m; T, H)$ as a function of the order parameter m. The minimum value of $f(m; T, H)$ at $m = m^*(T, H)$ gives $f(T, H)$. Therefore, we may regard $f(m; T, H)$ as a sort of potential function for the order parameter.

2.2.1 Self-consistent Equation of the Magnetization

Now, we review the phase transition of the ferromagnetic Ising model (2.1). In this system, the ground state is a complete ordered state ($\sigma_1 = \sigma_2 = \cdots = \sigma_N = \pm 1$), and thus the model is called ferromagnetic. The order parameter of the system is the magnetization:

$$M = \sum_{i=1}^{N} \sigma_i, \quad \text{or} \quad m = \frac{M}{N}. \tag{2.3}$$

By using standard mean-field analysis (see Appendix 8.1), the equation of state is obtained as the selfconsistent equation:

$$\tanh(\beta Jzm + \beta H) = m, \tag{2.4}$$

where z is the number of nearest neighbor sites, e.g., $z = 4$ for the square lattice. From this relation, the critical temperature T_C, below which nonzero solutions appear, is obtained as

$$T_C = \frac{Jz}{k_B}. \tag{2.5}$$

2.2.2 *Magnetization Curve*

In Fig. 2.2, the magnetization curves $M(H)$ for various temperatures are depicted. At the high temperatures (thin lines: $T = 1.2\ T_C$ and $1.5\ T_C$), the system is in a disordered state (paramagnetic state) and $m(H)$ is monotonic. The slope dm/dH at $H = 0$ is the susceptibility.

$$\chi = \left.\frac{dm(H)}{dh}\right|_{H=0}. \tag{2.6}$$

The slope of $m(H)$ at $H = 0$ increases when the temperature approaches T_C, and it diverges at T_C (dashed line). Below the critical temperature (bold line:$T = 0.7\ T_C$ and $0.8\ T_C$), $m(H)$ shows a hysteresis. In this case, the system has a spontaneous magnetization (noted by dots at B):

$$m_s = \lim_{H\to+0} m(H) > 0, \tag{2.7}$$

when the field is reduced from the point A to the point B. When the field approaches to 0 from negative side (B'), the system has the spontaneous magnetization $-m_s$. The state with positive magnetization $m(H)$ continues to the side of $H < 0$ (B-D). In the part (B-D), the system is metastable, and the endpoint D (noted by open circle) is the spinodal point where the metastability disappears and jumps to the stable state X. The part D-D' is thermodynamically unstable because

$$\left(\frac{\partial m}{\partial H}\right)_T < 0, \tag{2.8}$$

the state is unstable, which will be seem in the free energy land scape in the next subsection.

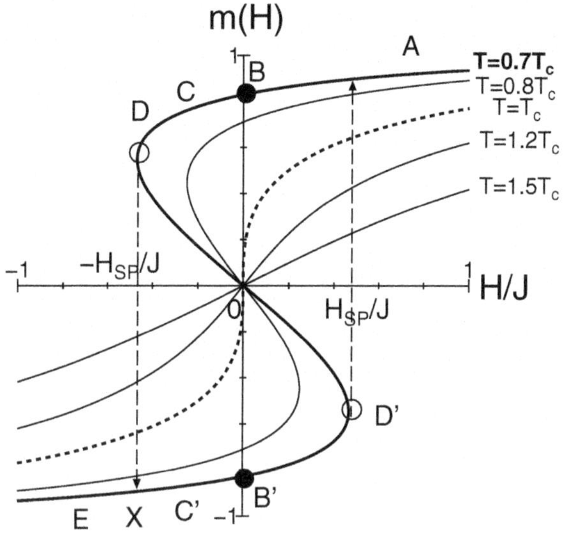

Fig. 2.2 Field dependences of the magnetization at various temperatures obtained by the mean-field analysis ($z = 4$) for $T = 0.7\ T_C, 0.8\ T_C, T_C, 1.2\ T_C$ and $1.5\ T_C$. At low temperatures ($T = 0.7\ T_C$: bold line, and $0.8\ T_C$), hysteresis appears. On the curve for $T = 0.7\ T_C$, typical positions of fields are noted. The metastable state survives until the spinodal point H_{SP}. The dashed line gives $m(H)$ at the critical point ($T = T_C$). The thin line gives a paramagnetic magnetization curve at a high temperature ($T = 1.2\ T_C$ and $1.5\ T_C$)

2.2.3 Free Energy as a Function of Magnetization

The situation of stability of state with m is well described by the phenomenological free energy $f(m : T, H)$ which is a kind of potential energy for the order parameter as the Landau free energy in the Landau theory of phase transition (see Sect. 2.5). In Appendix 8.1, various ways to obtain the free energy as a function of magnetization are explained. Here, we adopt the form obtained by the so-called Brag-Williams approximation:

$$F_{\rm BW}(m) = -\frac{JzN}{2}m^2 - HNm + k_{\rm B}TN\left(\frac{1+m}{2}\ln\frac{1+m}{2} + \frac{1-m}{2}\ln\frac{1-m}{2}\right), \tag{2.9}$$

which is depicted in Fig. 2.3a. For a large positive field, e.g., at the point a in Fig. 2.2, $F_{\rm BW}(m : T, H)$ has a single minimum. When the field becomes 0 (the points B and B'), the system is in the coexistence state where the two minima of $F_{\rm BW}(m : T, 0)$ have the same value. As the field becomes negative, the left minimum in the negative field (point C') becomes smaller, but the right minimum in the positive field (point C) is still locally stable. Between them, a free energy barrier exists.

When the field further decreases, the left minimum becomes unstable (point D). This point is called 'spinodal point'. This point is given by the following conditions. Because the point is an extremum of the free energy (self-consistent equation):

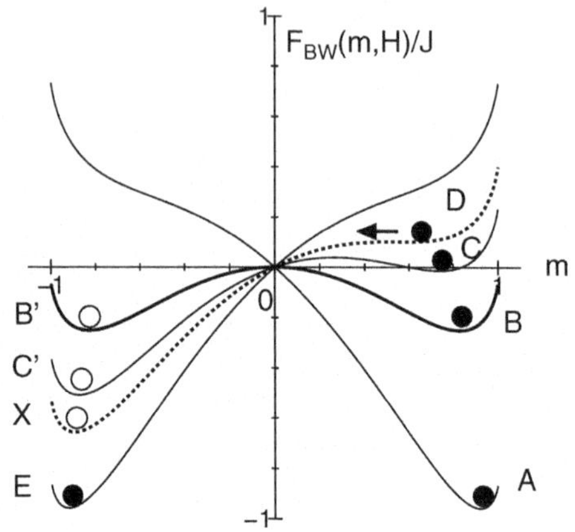

Fig. 2.3 The free energy as function of the magnetization at $T = 0.7\,T_C$: $F_{BW}(m, H) - F_{BW}(0, H)$. The points, A, B, C, D, B', C' X, E, correspond to those in Fig. 2.2 at which $H = 0.8J$, $H = 0$ (coexistent case, bold line), $H = -0.3J$, $H = H_{SP}$ (broken line), and $H = -0.8J$

$$\frac{\partial F(m, H, T)}{\partial m} = 0 \quad \rightarrow \quad m - \tanh(\beta J z m + \beta H) = 0, \tag{2.10}$$

and the condition that the point is at the border between upward-convex function and downward-convex function is

$$\frac{\partial^2 F(m, H, T)}{\partial m^2} = 0 \quad \rightarrow \quad 1 - \frac{\beta J z}{\cosh^2(\beta J z m + \beta H)} = 0. \tag{2.11}$$

From these equations, values of the field of the spinodal point at a temperature are given by

$$H_{SP} = \mp J z X \pm \frac{1}{2\beta} \ln\left(\frac{1+X}{1-X}\right) \quad X \equiv \sqrt{1 - \frac{1}{\beta J z}}, \tag{2.12}$$

which is depicted by the bold line in Fig. 2.4, inside of which the metastable state exists. At the spinodal point, the magnetization is given by

$$m_{SP} = \pm\sqrt{1 - \frac{1}{\beta J z}}. \tag{2.13}$$

In Fig. 2.4, the phase diagram and shapes of the free energy functions $f(m; T, H)$ in regions of the phase diagram are depicted.

It should be noted that the free energy barrier in the MF is macroscopic, i.e., proportional to the system size N. In the limit $N \rightarrow \infty$ the lifetime is infinity, and the system shows a criticality at the spinodal point. But, in finite size systems, the relaxation time is finite, which will be studied in the next chapter (see Sect. 3.4).

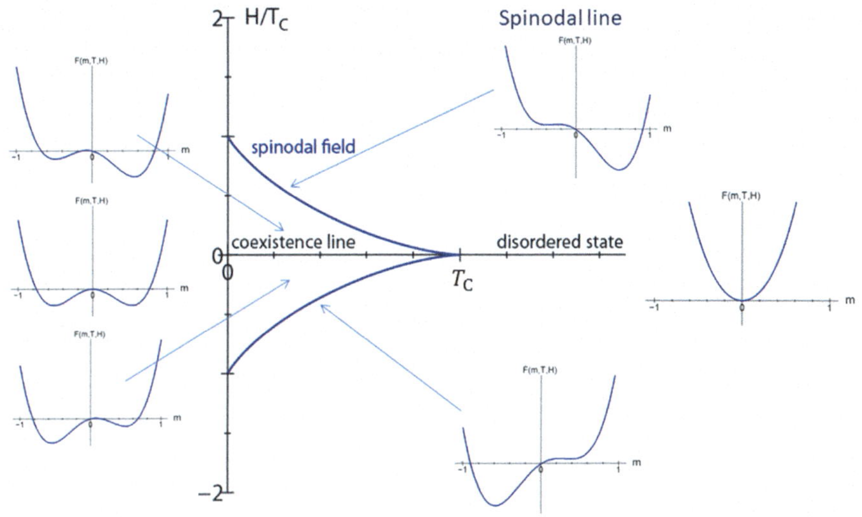

Fig. 2.4 Phase diagram of ferromagnetic model for the parameters (T, H) including the spinodal line (bold line). The line of $H = 0$ at $T \leq T_C$ denotes the coexistent state, at which the equilibrium magnetization changes discontinuously. Shapes of the free energy functions $f(m; T, H)$ in regions of the phase diagram are also shown

2.3 Rotation of Magnetization

So far, we studied the discontinuous change of the magnetization in the Ising model, where the magnetization is a scalar variable along the z-axis. In this case, the amplitude of the total magnetization must vanish ($m = 0$) at a point during magnetization reversal. On the other hand, magnets are generally modeled by a continuous spin system such as the anisotropic Heisenberg model in which the magnetization is given by a three-dimensional vector, i.e., $\boldsymbol{S}_i = (S_i^x, S_i^y, S_i^z)$, and the Hamiltonian is given by

$$\mathcal{H} = -J \sum_{<i,j>} \boldsymbol{S}_i \cdot \boldsymbol{S}_j - D \sum_{i=1}^{N} (S_i^z)^2 - \sum_{i=1}^{N} \boldsymbol{H} \cdot \boldsymbol{S}_i, \quad D > 0, \tag{2.14}$$

where D is the single-ion anisotropy. In this case, the magnetization is a vector and it can rotate to change the sign.

Because of the ferromagnetic interaction, spins in the system are tightly aligned at low temperatures. There, the dynamics of the state may be expressed as a motion of the total magnetization

$$\boldsymbol{m} = \frac{1}{N} \sum_{i}^{N} \boldsymbol{S}_i, \tag{2.15}$$

which can rotate keeping its amplitude, i.e., $|m|^2 = m_x^2 + m_y^2 + m_z^2 \simeq 1$. When the magnetization reverses in an antiparallel field, the rotation does not cause the breakdown of magnetization, i.e., $m_z = 0$, $|m| \simeq 1$, and the reversal can occur without loss of the exchange energy in contrast to the case of Ising model.

2.3.1 *Stoner-Wohlfarth Model*

If we assume the length of magnetization m to be a constant, we may model the system using a unit vector:

$$\mathcal{H} = -D(m_z)^2 - \boldsymbol{H} \cdot \boldsymbol{m}, \quad \boldsymbol{H} = (H_x, 0, H_z), \quad |\boldsymbol{m}| = 1, \tag{2.16}$$

where D is rescaled from that of (2.14) as $DN \to D$. Here, we set $H_y = 0$ and $m_y = 0$ without loss of generality. This model is called 'Stoner-Wohlfarth model [9].

First, we consider the case $H_x = 0$. The energy as a function of the field is simply given by

$$E(m_z) = -D(m_z)^2 - H_z m_z = -D\cos^2\theta - H_z \cos\theta. \tag{2.17}$$

The energies for various values of H are depicted as a function of θ in Fig. 2.5.

Let the initial state be a state with $\boldsymbol{m} = (0, 0, 1)$, which is a stable state in a positive magnetic field $\boldsymbol{H} = (0, 0, H_z > 0)$. For $H_z > 0$, the energy has a single minimum at $\boldsymbol{m} = (0, 0, 1)$ which is an absolute stable state. If H_0 decreases and reaches 0, then the energy has degenerate minima at $\boldsymbol{m} = (0, 0, \pm 1)$. When the field changes the sign to $H_z < 0$, the state $\boldsymbol{m} = (0, 0, -1)$ becomes the truely stable state. But, the state $\boldsymbol{m} = (0, 0, 1)$ is still locally stable, i.e., metastable, until a certain value of the field, which we call Stoner-Wohlfarth point

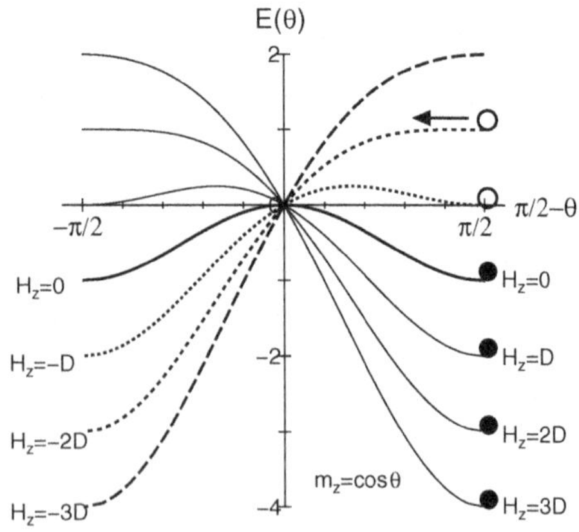

Fig. 2.5 Dependence of the energy of the Stoner-Wohlfarth model (2.17) on θ for various values of H

$$H_{\mathrm{SW}} = -2D. \tag{2.18}$$

At this point, the state $\boldsymbol{m} = (0, 0, 1)$ becomes unstable. This change of magnetization gives the simplest picture of coercivity of magnets.

2.3.2 Stoner-Wohlfarth Diagram

Next, we study the case of $H_x \neq 0$. That is, the magnetic field is not parallel to the easy axis, i.e., z axis. If $D = 0$, the stable direction is simply along the field $\boldsymbol{H}$. In contrast, when $D > 0$, the anisotropy energy causes an effective field given by

$$\boldsymbol{H}_{\mathrm{eff}} = \frac{d\mathcal{H}}{d\boldsymbol{m}} = \begin{pmatrix} H_x \\ 0 \\ H_z + 2Dm_z \end{pmatrix}, \quad \boldsymbol{H} = \begin{pmatrix} H_x \\ 0 \\ H_z \end{pmatrix}. \tag{2.19}$$

The stable direction for $H_x \neq 0$ is along $\boldsymbol{H}_{\mathrm{eff}}$ and no more parallel to $\boldsymbol{H}$. Then, dependence of the direction of the magnetization on the field becomes complicated as depicted in Fig. 2.6.

It is known that the point (H_x, H_y, H_z) at which the metastable state becomes unstable satisfies the relation [9]:

$$(2D)^{2/3} = (H_x)^{2/3} + (H_y)^{2/3} + (H_z)^{2/3}. \tag{2.20}$$

Thus, for a given value of H_x, the threshold field H_z^{SW} at which the metastable state becomes unstable is given by the relation

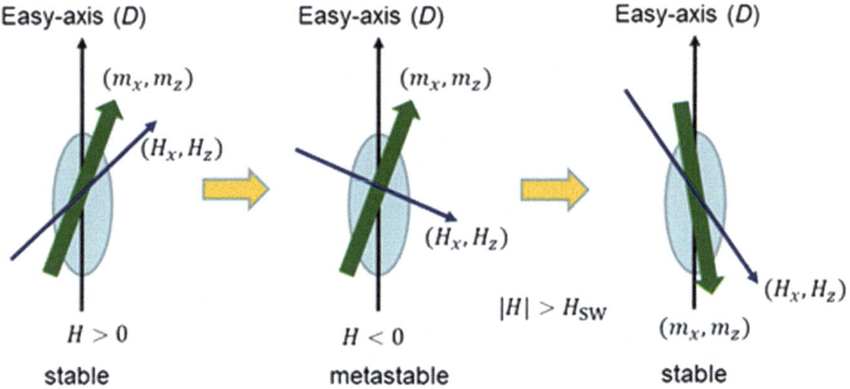

Fig. 2.6 Spin and field in Stoner-Wohlfarth model

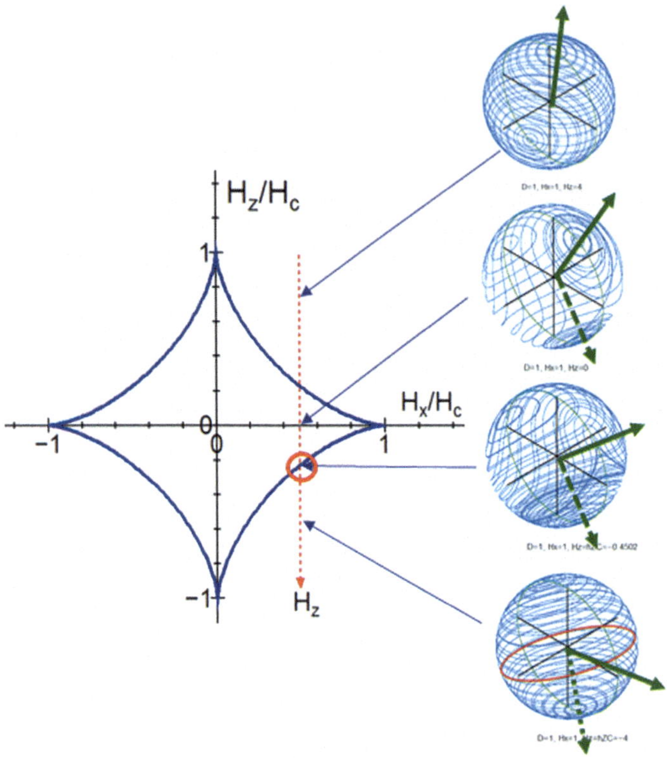

Fig. 2.7 Stoner–Wohlfarth diagram in the space of (H_x, H_z). The bold line gives (2.21) where H_c denotes $2D$. Curves on the spheres on the right side of the figure denote contours of same energy, i.e., trajectories of classical motion, which will be given in the following subsection (Fig. 2.6). The solid arrow denotes a local minimum direction of the magnetization. The dashed arrows denote the direction of negative magnetization with the minimum energy

$$H_z^{\mathrm{SW}}(H_x) = \left((2D)^{2/3} - (H_x)^{2/3}\right)^{3/2}. \tag{2.21}$$

A phase diagram in (H_x, H_z) plane of the point $H_z^{\mathrm{SW}}(H_x)$ which satisfies (2.21) is called 'Stoner–Wohlfarth diagram'. This is depicted in Fig. 2.7. This kind of phase diagram has been observed experimentally for single nano-size magnet by the micro-SQUID technique [10].

Let us consider the change of state when H_z moves along the dotted arrow in the phase diagram in the figure. In the right of the figure, we depict the direction of positive and negative stable magnetizations by solid and dotted arrows, respectively. On spheres, trajectories of magnetizations are plotted which will be explained in the next subsection.

At the top of dotted arrow in the phase diagram, the system has a single energy minimum point where $m_z > 0$. When H_z reaches the upper curve of $H_z^{\mathrm{SW}}(H_x)$ in the phase diagram, a metastable state with a negative magnetization (dashed arrow) appears. But it has no effect on the stable state of positive magnetization. When H_z reaches the horizontal line $H_z = 0$, the system has two minima of positive and negative magnetizations $m_z = \pm m_0$:

$$m_0 = \sqrt{1 - \left(\frac{H_x}{2D}\right)^2}. \tag{2.22}$$

That is, the system is at the first-order phase transition point, and both of the stable points have the same energy (the second sphere). When H_z becomes negative, the state with negative magnetization becomes the true stable state. But the state with positive magnetization still metastable. When H_z reaches the lower bold curve plotting $-H_z^{\mathrm{SW}}(H_x)$, the state of positive magnetization becomes unstable (the third sphere). When the field changes to further negative value, the system has a unique stable point with $m_z < 0$. Thus, the metastable magnetization cannot continue over $H = H_{\mathrm{SW}}$, and thus it shows a discontinuous change. Because of the energy conservation, the magnetization does not relax to the stable state (the dashed arrow), and the magnetization of the solid arrow shows a rotation around the solid circle (the bottom sphere).

2.3.3 *Trajectrory*

In classical systems, the magnetization deterministically moves with the torque equation

$$\frac{d\boldsymbol{m}}{dt} = \boldsymbol{m} \times \boldsymbol{H} = \boldsymbol{m} \times \begin{pmatrix} H_x \\ 0 \\ H_z + 2Dm_z \end{pmatrix}, \tag{2.23}$$

which draws a trajectory as an equi-energy curve.

In Fig. 2.8, we depict trajectories with various energies for several values of H_z. When the magnetic field is swept slowly from a large positive field, the magnetization $m(t)$ starting from the stable state moves adiabatically until the field reaches to $-H_{\mathrm{SW}}$. The motion of the magnetization under the sweep is shown by dots in the spheres.

The bottom left sphere shows trajectories at $H_z = -H_z^{\mathrm{SW}}(H_x)$, where the trajectory shows a singular shape reflecting infinite relaxation time of $m(t)$ at the trajectory pointed by the bold arrow. When the field exceeds $H_z = -H_z^{\mathrm{SW}}(H_x)$, $m(t)$ rotates around $\boldsymbol{H}_{\mathrm{eff}}$ as the trajectory pointed by the bold arrow in bottom middle sphere.

In the bottom right of the figure, the motion of magnetization $\boldsymbol{m}(t)$ under a slowly swept field is shown. The magnetization follows the local minimum point (shown by bullets on spheres) almost adiabatically until $H_{\mathrm{SW}}(H_x)$. But, after $H_{\mathrm{SW}}(H_x)$, $m(t)$ rotates around $\boldsymbol{H}_{\mathrm{eff}}$. In a dissipative environment, the system loses the emergy and the magnetization relaxes to the stable point.

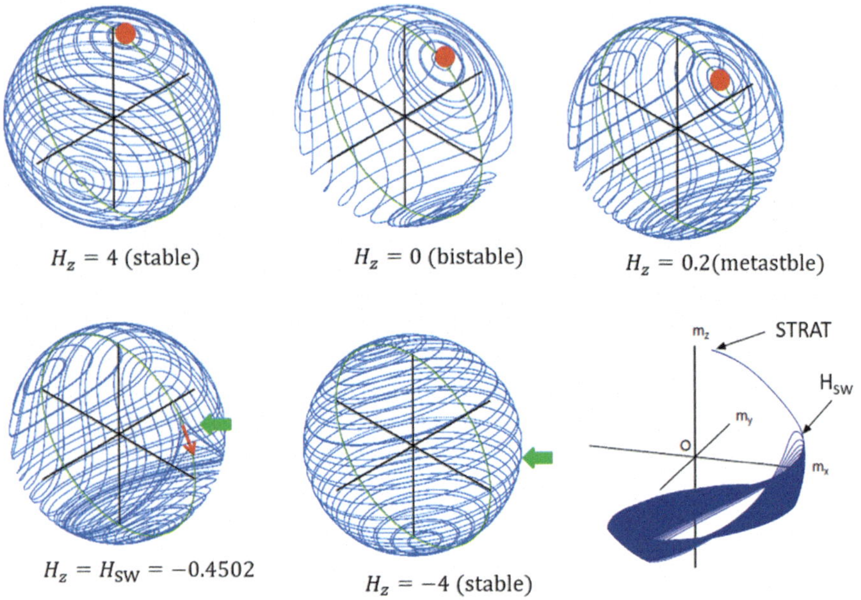

Fig. 2.8 Trajectories in the classical model. $D = 1$, $H_x = 1$ at several values of H_z

2.4 First-Order Phase Transitions as a Function of the Temperature

So far, we studied the phase transition as a function of the external field. The first-order phase transition also occurs as a function of the temperature. In this section, we introduce some examples of such models.

2.4.1 A Model with Many Degeneracy of Zero Energy States

In the Ising model, the variable takes two values $\sigma = \pm 1$. Now, we consider a model in which the variable takes values $S = -1, 0$ and 1, which is the Ising model of $S = 1$. This model still exhibits the second-order phase transition. However, when the number of states of $S = 0$ increases, the model exhibits the first-order phase transition.

Let us consider the following model

$$\mathcal{H} = -J \sum_{<ij>} S_i S_j - H \sum_i S_i, \quad S_i = -1, \underbrace{0, 0, \ldots, 0}_{n_0 \text{fold}}, 1. \tag{2.24}$$

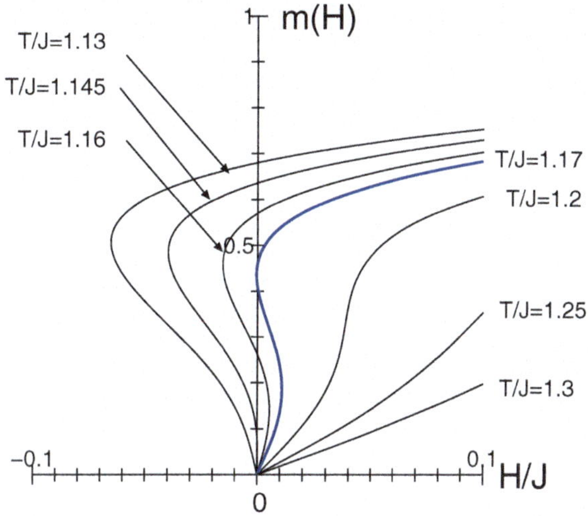

Fig. 2.9 The temperature dependence of the magnetization curve $m(H)$ of the model (2.25). $n = 5$ and $z = 4$

Let numbers of states which in which $S = (-1, 0, 1)$ be (n_{-1}, n_0, n_1), respectively. Here, we study the case of $(n_{-1}, n_0, n_1) = (1, n, 1)$. Analyzing the model by a mean-field theory, the self-consistent equation is given by

$$m = \frac{e^{\beta Jzm+\beta H} - e^{-\beta Jzm-\beta H}}{e^{\beta Jzm+\beta H} + n + e^{-\beta Jzm-\beta H}}. \tag{2.25}$$

This is rewritten as

$$H(m) = -Jzm + k_{\mathrm{B}}T \ln\left(\frac{nm + \sqrt{(nm)^2 + 4(1-m^2)}}{2(1-m)}\right), \tag{2.26}$$

and $H(m)$ for $n = 5$ and $z = 4$ is depicted in Fig. 2.9 for several temperatures. In the figure, only the part of $H \geq 0$ are shown because $m(-H) = -m(H)$. At the high temperatures $T = 1.3$, 1.25 and 1.2, the system is paramagnetic, that is, the magnetization monotonically increases with the field. But, at lower temperatures, non-monotonic behavior appears. Around $T = 1.17$, a solution with nonzero m at $H = 0$ appears. Thus, there are solutions $m = 0$ and $m = \pm m_0$. Here m_0 does not start from zero continuously, but it appears discontinuously, while, in the second order phase transition, non-zero solution $\pm m_0$ starts from zero continuously as we saw in the standard Ising model. That is, here, a first-order phase transition occurs. At $T = 1.145$, dm/dH at $H = 0$ diverges, which indicates that the state of $m = 0$ becomes unstable. At the low temperature $T = 1.13$, $m = \pm m_0$ are the stable states, and $m = 0$ is unstable.

The first-order phase transition occurs owing to large degeneracy of the $S = 0$ state, i.e., an entropy effect. This type of mechanism of the first-order phase transition has been studied also in the Potts model [11].

2.4.2 Blume-Capel Model

The same scenario works in the Blume-Capel model. The Blume-Capel model is a model of $S = 1$ Ising model with an in-plane anisotropy term [12]:

$$\mathcal{H}_{\rm BC} = -J\sum_{<ij>} S_i S_j + D\sum_{i=1}^{N}(S_i)^2, \quad S_i = -1, 0, 1, \quad D > 0. \tag{2.27}$$

Here, the degeneracy of states is $(n_{-1}, n_0, n_1) = (1, 1, 1)$. The self-consistent equation of the model is

$$m = \frac{e^{\beta(Jzm-D)} - e^{-\beta(Jzm+D)}}{e^{\beta(Jzm-D)} + 1 + e^{-\beta(Jzm+D)}} = \frac{e^{\beta Jzm} - e^{-\beta Jzm}}{e^{\beta Jzm} + e^{\beta D} + e^{-\beta Jzm}}. \tag{2.28}$$

In this model, although the degeneracy n_0 is 1, the thermal weight of the state of $S_i = 0$ is larger than those of $S_i = \pm 1$ because of the term of D. Therefore, the model with a large value of D exhibits the first-order phase transition in a similar way to that in the previous subsection. For small values of D, the model shows the second-order phase transition. The border between the part of the second-order phase transition and that of the first-order phase transition is called 'tricritical point'.

2.4.3 Spin Crossover Systems

Another mechanism of the first-order phase transition as a function of the temperature is found in the spin crossover model [13], which is a model for electron states of an atom or ion. For example, spins of electrons at the 3d orbit of an atom tend to align by the Hund rule and the total spin has a large value, e.g., in Fe^{3+} ion, the 3d orbit has 5 electrons and the spins form a composite spin of $S = 5/2$. This state is called 'high-spin state' (HS). However, in a crystal, the ligand (crystal) field may cause a split the 3d orbit to the t_{2g} and e_g orbits. If the split is large, the electrons tend to fill the low-energy states. In each energy level, electrons with $\pm$ spins are filled because of the Pauli principle. Thus, the total spin is $S = 1/2$, which is called 'low-spin state' (LS). The two cases are depicted in Fig. 2.10.

Let the energies of HS state and LS state be $E_{\rm HS}$ and $E_{\rm LS}$, respectively, and also the degeneracies be $n_{\rm HS}$ and $n_{\rm LS}$ for them. Usually, HS state has a higher energy and a larger degeneracy than those of LS state, i.e.,

$$E_{\rm HS} > E_{\rm LS}, \quad n_{\rm HS} > n_{\rm LS}. \tag{2.29}$$

If the difference of energies of HS and LS states is not large, a transition between HS and LS states can be induced by effects of temperature, pressure, and light-irradiation, etc., which is called 'spin-crossover transition' [13–15]. The change between HS and

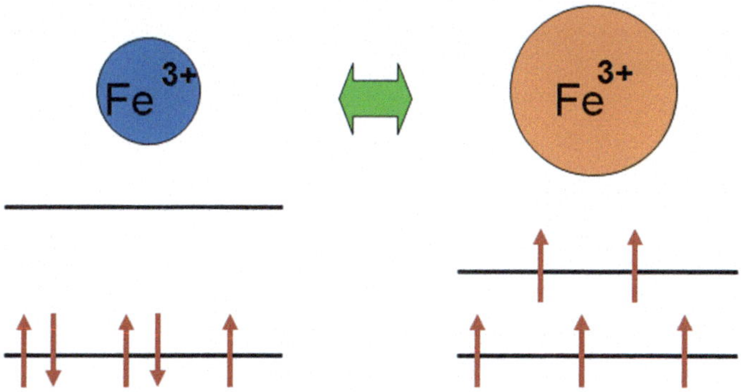

Fig. 2.10 LS state (left) and HS state (right) of Fe^{3+}

LS states may occur smoothly or discontinuously depending on situations, which we will study below. To model such phenomena, an Ising spin has been introduced to express HS state and LS state by $S_i = \pm 1$, respectively [14].

To express the cooperativity among the spin states, a ferromagnetic interaction between them is assumed, which causes the neighboring states to be the same. Here, the model is not a magnetic system, but we still call the interaction 'ferromagnetic' in the terminology of the Ising model. The Hamiltonian of the system is given in the form:

$$\mathcal{H}_{\rm SC} = -J \sum_{<ij>} S_i S_j - H \sum_{i=1}^{N} S_i, \quad S_i = \underbrace{-1, -1 \ldots, -1}_{n_{\rm LS}\rm fold}, \underbrace{1, 1, \ldots, 1}_{n_{\rm HS}\rm fold}, \tag{2.30}$$

where the field H expresses the difference of $E_{\rm HS}$ and $E_{\rm LS}$ as

$$H = -D, \quad D = \frac{E_{\rm HS} - E_{\rm LS}}{2}, \quad E_0 = \frac{E_{\rm LS} + E_{\rm HS}}{2} \tag{2.31}$$

and

$$E_{\rm LS} = E_0 + H \quad E_{\rm HS} = E_0 - H. \tag{2.32}$$

Hereafter, we set $E_0 = 0$ in the Hamiltonian because it is a constant.

The self-consistent equation is given by

$$m = \frac{n_{\rm HS} e^{\beta J z m + \beta H} - n_{\rm LS} e^{-\beta J z m - \beta H}}{n_{\rm HS} e^{\beta J z m + \beta H} + n_{\rm LS} e^{-\beta J z m - \beta H}}$$

$$= \frac{\sqrt{n_{\rm HS} n_{\rm LS}} (e^{\beta J z m + \beta H} \sqrt{\frac{n_{\rm HS}}{n_{\rm LS}}} - e^{-\beta J z m - \beta H} \sqrt{\frac{n_{\rm LS}}{n_{\rm HS}}})}{\sqrt{n_{\rm HS} n_{\rm LS}} (e^{\beta J z m + \beta H} \sqrt{\frac{n_{\rm HS}}{n_{\rm LS}}} + e^{-\beta J z m - \beta H} \sqrt{\frac{n_{\rm LS}}{n_{\rm HS}}})}. \tag{2.33}$$

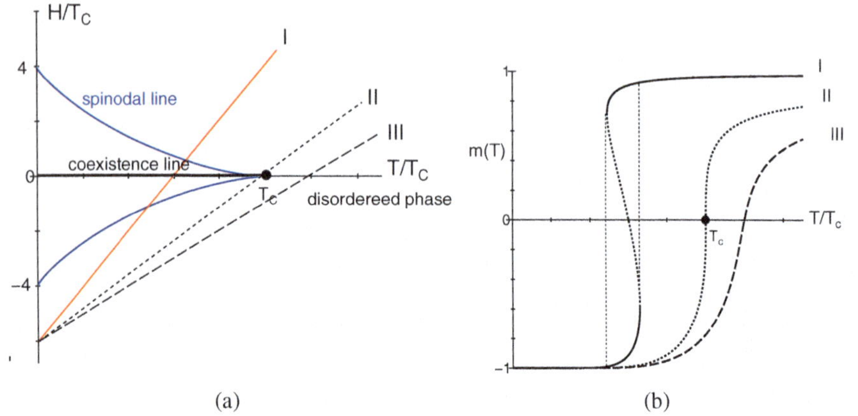

Fig. 2.11 **a** Phase diagram of the Ising model in the space (T, H). The curves show the spinodal line inside which the system has metastable state. The bold line denotes the coexistence line where two states with opposite magnetizations coexist. Three types of temperature dependence of states of the spin-crossover system are shown in the phase diagram of the Ising model. (I) a case with a first-order phase transition, (II) a case with a second-order phase transition, and (III) a case of continuous transition. **b** Corresponding temperature dependence of the order parameter

By setting $g = n_{\mathrm{HS}}/n_{\mathrm{LS}}$, i.e., $\sqrt{n_{\mathrm{HS}}/n_{\mathrm{LS}}} = (1/2)\ln g$, it is rewritten as

$$m = \frac{e^{\beta Jzm+\beta H+\frac{1}{2}\ln g} - e^{-\beta Jzm-\beta H-\frac{1}{2}\ln g}}{e^{\beta Jzm+\beta H+\frac{1}{2}\ln g} + e^{-\beta Jzm-\beta H-\frac{1}{2}\ln g}}. \tag{2.34}$$

This form can be regarded as the self-consistent equation of a simple Ising model with $\sigma = \pm 1$ in the effective field H_{eff}.

$$\mathcal{H}_{\mathrm{SC-Ising}} = -J\sum_{<ij>}\sigma_i\sigma_j - H_{\mathrm{eff}}\sum_{i=1}^{N}\sigma_i, \quad \sigma_i = -1, 1, \tag{2.35}$$

where

$$H_{\mathrm{eff}} = -D + \frac{1}{2}k_{\mathrm{B}}T\ln g. \tag{2.36}$$

Thus, a state in the SC system at T and H is given by a point $(T, H = -D + \frac{1}{2}k_{\mathrm{B}}T\ln g)$ in the phase diagram of the Ising model (Fig. 2.4). Thus, depending on the value of D and g, various types of temperature dependence of $\langle\sigma\rangle$ are realized as depicted in Fig. 2.11a.

In case I, the line of $(T, H = -D + \frac{1}{2}k_{\mathrm{B}}T\ln g)$ crosses the coexistence line, i.e., $H = 0$ at $T < T_{\mathrm{C}}$. At this point, the equilibrium magnetization changes discontinuously. When the field is swept, metastable state survives until the line crosses with the spinodal line. Thus, the temperature dependence of $\langle\sigma\rangle$ shows a hysteresis which is depicted in Fig. 2.11b. In case III, the dotted line crosses the line of $H = 0$ at

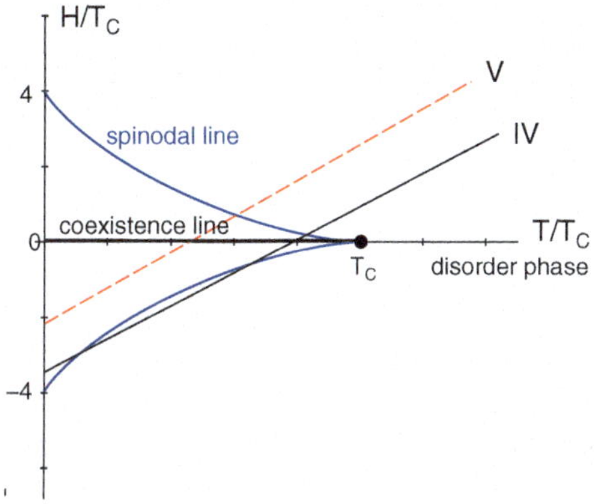

Fig. 2.12 Phase diagram of the Ising model in the space (T, H). The curves show the spinodal line inside which the system has metastable state. The bold line denotes the coexistence line where two states with opposite magnetizations coexist. Two types of temperature dependence are shown in the phase diagram of the Ising model. (IV) a case with a first-order phase transition and a HS metastable state at low temperature, and (V) a case with HS state at all the temperature with a hidden LS (stable) state at low temperatures

$T > T_C$, and thus $\langle\sigma\rangle$ changes continuously, which is also shown in Fig. 2.11b. In case II, the dotted line crosses at the critical point of $H = 0$. This gives a special case of the second-order phase transition. There, the slope of $\langle\sigma\rangle$ diverges at $T = T_C$.

There are some peculiar metastable structures. In Fig. 2.12, we show cases with a reentrant type metastable HS state (type IV) and also a hidden metastable HS state at low temperatures (type V). In case IV, a first-order phase transition between HS and LS states occurs at high temperature. When the line goes outside of the metastable region, LS state is singly stable and HS state is no more metastable. However, when the line comes back into the metastable region at low temperature, HS metastable state appears again.

In spin-crossover materials, so-called light-induced excited spin state trap (LIESST) [16] attracts interests. When a light of certain frequency is irradiated on the equilibrium LS state at a low temperature, the spin state is excited to HS state. LIESST is a phenomenon that the induced HS state survives for a long time. This long survival time may be originated from two mechanisms. One of them is an atomic metastability. That is, reorganization of state takes a long time due to slowdown of microscopic processes which obey the Arrhenius law. The other is a thermodynamic metastability. That is, the low-temperature metastable state found in the process IV in Fig. 2.12 keeps the HS state as a metastable state. In the former case, HS fraction decreases monotonically after the irradiation, while in the latter case, it moves to the HS metastable value in a short time and it relaxes to the stable LS state much later [17]. Trap of HS state was experimentally found by quick quench of the temperatures, and the existence of the low-temperature metastable state was observed [18].

In case V, the line is always in the metastable region at low temperature. Thus, when the system is cooled down the system remains in HS state, although LS state is true equilibrium state for $H < 0$. In this case, to bring the system to the LS state we need some disturbance to make the system jump over the barrier. The process of case V was also found experimentally, where a strong photo-irradiation caused the transition from the metastable HS state to the equilibrium LS state [19].

Moreover, because the sizes of molecule in HS and LS states are different, the interaction among spins becomes long-range due to the elastic interaction of lattice distortion. This long-range ferromagnetic interaction causes the phase transition of the system to be mean-field type [20–23]. This property causes, besides the above-mentioned cases, more complicated ordering processes. In particular, in systems with antiferromagnetic nearest neighbor interactions, complicated phase diagrams appear due to the competition between the long-range ferromagnetic interaction and the short-range antiferromagnetic interaction [24].

2.5 Landau Theory

We have studied the mechanism of phase transitions by using the free energy as a function of order parameter $f(m; T, H)$ obtained by the mean-field theory, e.g., (2.9). Although the form of $f(m; T, H)$ depends on details of the models, properties of phase transition qualitatively have the same structure, which can be captured by a simplified polynomial form of $f(m; T, H)$, which is called 'Landau phenomenological free energy' or simply 'Landau free energy'. For example, if we expand $f(m; T, H)$ given in (2.9) as a function of a polynomial of m, we have

$$f(m; T, H) = a(T)m^2 + b(T)m^4 + \cdots - Hm, \tag{2.37}$$

where

$$a = \frac{\beta Jz}{2}(k_B T - Jz), \quad b = \frac{\beta Jz}{12}(\beta Jz)^3. \tag{2.38}$$

This simple form can reproduce nature of the phase transition well. For example, the second-order phase transition at $H = 0$ is characterized by the change of the sign of a at T_C. Thus, we express it as

$$a(T) = a_0(T - T_C), \quad a_0 > 0. \tag{2.39}$$

Keeping b to be positive, the free energy has the form

$$f(m; T, H) = a_0(T - T_C)m^2 + bm^4 - Hm, \tag{2.40}$$

and thus the magnetization at a temperature T is given by

$$\frac{df(m;T,H)}{dm} = 0 \to 2a_0(T - T_C)m + 4bm^3 - H = 0, \tag{2.41}$$

which corresponds to an expansion of the self-consistent equation $m = \tanh(\beta Jzm + \beta H)$ and gives all the critical behavior qualitatively. For example, at $H = 0$, the spontaneous magnetization m_s appears below the critical temperature T_C as

$$m_s = \pm\sqrt{\frac{a_0(T_C - T)}{2b}}, \tag{2.42}$$

while $m = 0$ is only solution above T_C. And the susceptibility

$$\chi = \left.\frac{dm}{dH}\right|_{H=0} \tag{2.43}$$

diverges when the temperature approaches to T_C. At $T > T_C$ and $T < T_C$, χ is given as

$$\chi_{T>T_C} = \frac{1}{2a_0(T - T_C)}, \quad \text{and} \quad \chi_{T<T_C} = \frac{1}{4a_0(T_C - T)}, \tag{2.44}$$

respectively. Therefore, (2.40) is enough to study the qualitative properties of the second-order phase transition.

As studied in Fig. 2.3b, the system exhibits a first-order phase transition when the field changes the sign below T_C. This transition is understood as a process that the double well potential of $f(m; T, H)$ at $H = 0$ is tilted by the field. This process is also reproduced by the Landau theory with changing H in the form (2.40) with a negative a.

$$f(m;T,H) = am^2 + bm^4 + \cdots - Hm, \quad a < 0, \quad b > 0. \tag{2.45}$$

When $H = 0$ the $f(m; T, H)$ has degenerate two minima. That is, $H = 0$ is the first-order phase transition point. When $H > 0$, $f(m; T, H)$ tilts and the extremum of a positive magnetization becomes lower, and the other becomes metastable. When H reaches the spinodal field, the local minimum in the negative magnetization disappears. The spinodal point is given by the conditions:

$$\frac{df(m;T,H)}{dm} = -2|a|m + 4bm^3 - H = 0, \tag{2.46}$$

and

$$\frac{d^2f(m;T,H)}{dm^2} = -2|a| + 12bm^2 = 0, \tag{2.47}$$

which leads

$$H_{sp} = b\left(\frac{2|a|}{3b}\right)^{3/2}. \tag{2.48}$$

2.5.1 Landau Theory for Temperature Induced First-Order Phase Transition

As we studied in Sect. 2.4, there are many systems in which first-order phase transition occurs as a function of the temperature. In the second-order phase transition, the spontaneous magnetization m_S appears continuously from 0 at the critical point. However, the spontaneous magnetization m_S appears discontinuously with a jump from 0 to a nonzero value as we studied in Sect. 2.4.

In this case, the free energy is symmetric for the sign of order parameter m in contrast to the above-mentioned first-order phase transition induced by the field. That is, the free energy is an even function of m. Therefore, there appear two nonzero solutions, i.e., $m = \pm m_0$ at the transition temperature, and the system must choose one of them by the symmetry breaking procedure. To express mechanism of this type of first-order phase transition in the Landau theory, we need to include terms up to the sixth order.

$$f(m) = am^2 + bm^4 + cm^6, \tag{2.49}$$

with coefficients $a > 0$, and $c > 0$. Here, the first-order phase transition occurs as a function of b. If $b > 0$, $m = 0$ gives the minimum of the free energy. To have the first-order phase transition from $m = 0$ to a finite value m_0, we explore the case with $b < 0$.

Now, we study shapes of the free energy by changing the coefficient b, keeping a and c to be positive numbers. In the case negative b, $f(m)$ may have five extrema as depicted in Fig. 2.13.

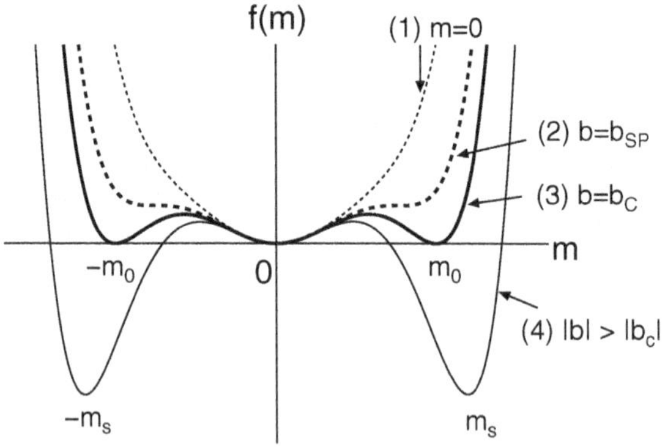

Fig. 2.13 Dependence of $f(m)$ on $b < 0$ for $a = 1, c = 1$: (1) at $b = -\sqrt{ac}$, $f(m)$ has single minimum at $m = 0$ (dotted curve). (2) at $b = b_{SP} = -\sqrt{3ac}$, $f(m)$ has a shoulder of the spinodal point (short-dash curve). (3) at $b = b_c = -2\sqrt{ac}$, $f(m)$ has degenerate extrema (bold curve). This is the first-order phase transition point. (4) at $b = -2.5\sqrt{ac}$, $f(m)$ has stable states with nonzero magnetization $\pm m_s$ (solid curve)

The extrema are given by

$$\frac{df(m)}{dm} = 2(a + 2bm^2 + 3cm^4)m = 0. \tag{2.50}$$

The solutions are

$$m = 0, \tag{2.51}$$

$$\pm m_{\rm us} = \pm\frac{-b - \sqrt{b^2 - 3ac}}{3c} = \pm\frac{|b| - \sqrt{b^2 - 3ac}}{3c}, \tag{2.52}$$

and

$$\pm m_{\rm s} = \pm\frac{-b + \sqrt{b^2 - 3ac}}{3c} = \pm\frac{|b| + \sqrt{b^2 - 3ac}}{3c}. \tag{2.53}$$

When $b^2 - 3ac < 0$, there is only one real solution $m = 0$. Thus we study the cases of $b^2 > 3ac$, i.e., $b < -\sqrt{3ac}$. From the figure, it is obvious that the solutions $\pm m_{\rm us}$ give unstable extrema. The solutions $\pm m_{\rm s}$ give metastable or stable solutions.

First, we consider the case of large negative b, e.g., $b = -2.5\sqrt{ac}$. As shown in Fig. 2.13, the free energy $f(m)$ has minima at nonzero $\pm m_0 = \pm m_{\rm s}$. When b increases ($|b|$ decreases), $f(m_s)$ increases, and the first-order phase transition occurs, at which $f(m_{\rm s})$ and $f(0)$ have the same value:

$$f(0) = 0 = f(\pm m_{\rm s}) = am^2 + bm^4 + cm^6 = 0. \tag{2.54}$$

Solving this relation with the condition (2.50), we have

$$b_{\rm c} = -2\sqrt{ac}. \tag{2.55}$$

When b increases further, $f(m_{\rm s})$ becomes larger than $f(0)$, and the $m = \pm m_{\rm s}$ become metastable as we see in Fig. 2.13 At $b^2 = 3ac$,

$$m_{\rm us} = m_{\rm s}, \tag{2.56}$$

and the local minimum at $\pm m_{\rm s}$ disappears, These give the spinodal point:

$$b_{\rm SP} = -\sqrt{3ac}. \tag{2.57}$$

In the present model, the local minimum at $m = 0$ does not become unstable because we fixed $a > 0$. To make the solution $m = 0$ to be unstable, we have to control a. At $a = 0$ it becomes unstable.

In Fig. 2.14, phase diagram in the space (a, b) is depicted. The line of $a = 0$ and $b > 0$ gives the second-order phase transition, the right of which is a region of disordered state, and the left of which is a region of ordered state with double-well shape of the free energy. The line of $a = 0$ and $b < 0$ gives the spinodal line at which

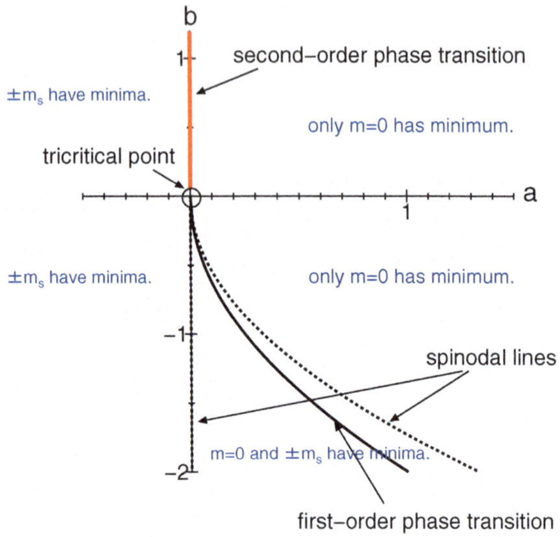

Fig. 2.14 Phase diagram in $a, b, (c = 1)$

the state of $m = 0$ becomes unstable. The spinodal line at which the state of $m = m_s$ becomes unstable is given by $b_{\text{spinodal}} = -\sqrt{3ac}$ (2.57) which is plotted by the dotted curve. In the region inside of the spinodal lines, the system has a metastable state. The system with $b < 0$ shows a first-order phase transition. The bold solid line gives the first-order transition point $b_{\text{spinodal}} = -\sqrt{2ac}$ (2.55). The point $(a = 0, b = 0)$ is called 'tri-critical point' which is the border between the first- and second-order phase transitions.

2.6 Gas-Liquid Phase Transition

Now, we refer to the Gas-liquid phase transition which is one of the most familiar phase transitions in the daily life. As we know, the ideal gas follows the Boyle-Charles law

$$PV = nRT, \tag{2.58}$$

where R is the gas constant ($= N_{\text{av}}k_{\text{B}}$) and N_{av} is the Avogadro number, and n is the number of particles N in mol, i.e., $N = n \times N_{\text{av}}$:

$$R = 8.314\,\text{JK}^{-1}\text{mol}^{-1}, \quad k_{\text{B}} = 1.38026 \times 10^{-23}\text{JK}^{-1}, \quad N_{\text{av}} = 6.02544 \times 10^{23}. \tag{2.59}$$

This relation does not show the gas–liquid phase transition.

To study the gas–liquid phase transition, we must take into account the interaction between particles $\phi(r)$. The Hamiltonian of the system with the interaction is given by

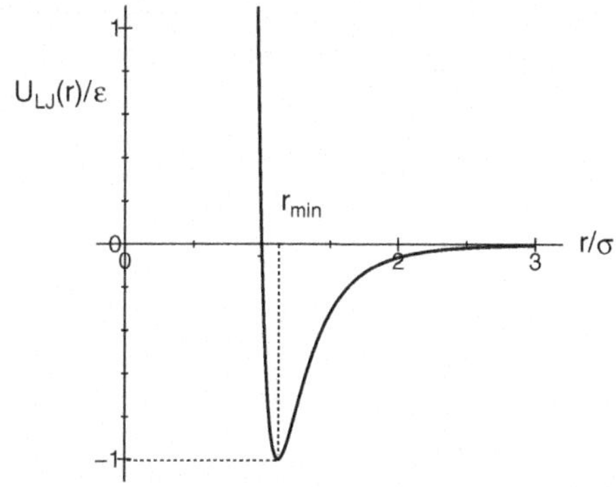

Fig. 2.15 Lennard-Jones potential

$$\mathcal{H} = \frac{1}{2m}\sum_{i=1}^{N} \boldsymbol{p}_i^2 + \sum_{i,j} \phi(|\boldsymbol{r}_i - \boldsymbol{r}_j|). \tag{2.60}$$

To explain the gas–liquid transition, there are two important ingredients: (1) effect of the excluded volume due to the finite size of particles and (2) a short-range interaction between particles.

The interaction between particles is often given by the Lennard-Jones potential.

$$U_{\mathrm{LJ}}(\boldsymbol{r}) = 4\varepsilon\left[\left(\frac{\sigma}{r}\right)^{12} - \left(\frac{\sigma}{r}\right)^{6}\right] \tag{2.61}$$

which is depicted in Fig. 2.15. The term $(\sigma/r)^6$ represents the van der Waals force which is an attractive interaction due to fluctuation of dipole-dipole interaction, and the term $(\sigma/r)^{12}$ represents the repulsive force due to cores. The minimum point of the potential is at $r_{\min} = 2^{1/6}\sigma$ where $U(r_{\min}) = -\varepsilon$. When two particles are bounded with this distance, the energy is this minimum value. If the particles come closer, the energy increases very rapidly, and thus σ represents twice of the core size.

2.6.1 Phenomenological Method: van der Waals (vDW) Equation

Before we analyze the model (2.60) by the statistical mechanics, we briefly review the van der Waals (vDW) equation which captures the essential properties of the gas-liquid transition. The vDW equation cleverly takes into account the ingredients (1) and (2) mentioned above. Due to the excluded volume effect (1), the volume of the system is effectively reduced as

$$V \to V - bN, \tag{2.62}$$

where b is a constant representing the size of particle, and the pressure is increased due to the attractive force between the particles (2) as

$$P \to P + a\left(\frac{N}{V}\right)^2 \tag{2.63}$$

where a is a constant representing strength of the interaction. Here it should be noted that pressure is defined as a force pushing the system from outside and the attractive force effectively increases it.

Thus, the equation of state of the ideal gas (2.58) is modified as

$$\left(P + a\left(\frac{N}{V}\right)^2\right)(V - bN) = nRT. \tag{2.64}$$

The coefficients a and b are called van der Waals constants which depend on the particles [25].

The vDW equation well describes many properties of real gases. At high temperatures, the volume dependence of the pressure $P = P(V)$ given by (2.64) is a monotonic decreasing function. But at low temperatures, it becomes non-monotonic. Because the thermodynamical stability requires the relation

$$\left(\frac{\partial P}{\partial V}\right)_T < 0, \tag{2.65}$$

the part of $\left(\frac{\partial P}{\partial V}\right)_T > 0$ is not allowed. There, the system must exhibit a phase transition.

Let us define the critical point T_{C} as the temperature at which $P = P(V)$ begins to be non-monotonic, i.e.,

$$\left(\frac{\partial P}{\partial V}\right)_{T_{\mathrm{C}}} = 0, \quad \left(\frac{\partial^2 P}{\partial V^2}\right)_{T_{\mathrm{C}}} = 0. \tag{2.66}$$

These relations give

$$V_{\mathrm{C}} = 3bN, \quad P_{\mathrm{C}} = \frac{a}{27b^2}, \quad T_{\mathrm{C}} = \frac{8a}{27k_{\mathrm{B}}b}. \tag{2.67}$$

With the scaled quantities

$$p = \frac{P}{P_{\mathrm{C}}}, \quad v = \frac{V}{V_{\mathrm{C}}}, \quad t = \frac{T}{T_{\mathrm{C}}}, \tag{2.68}$$

the van der Walls equation (2.64) is rewritten as as

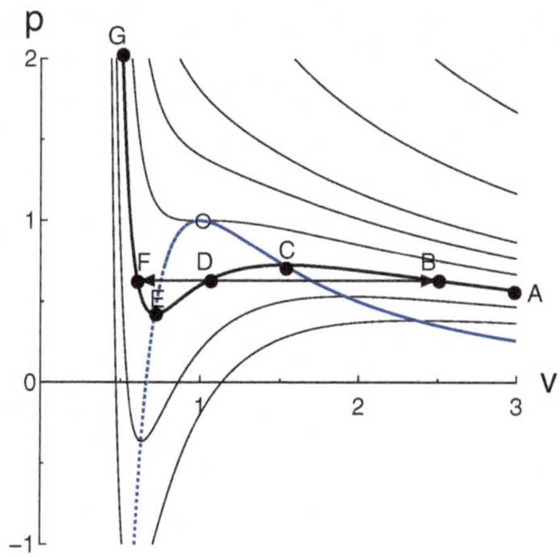

Fig. 2.16 PV relations given by the van der Waals equation (2.64). Temperatures are $T/T_C =$ 0.7, 0.8.0.9, 1.0, 1.1, 1.2, 1.5, and, 2.0. The white circle shows the critical point. (The bold curve depicts the relation at $T/T_C = 0.9$.) State is thermodynamically unstable ($(\partial P/\partial V)_T > 0$) inside the dotted line which gives the spinodal line

$$\left(p + \frac{3}{v^2}\right)(3v - 1) = 8t. \tag{2.69}$$

The parameters a and b disappears in this equation, which means that all the gases with various values of a and b are commonly expressed by this equation, and the states having the same scaled quantities (2.68) is called 'corresponding state'.

The temperature dependence of (2.69) is depicted in Fig. 2.16. As mentioned above, at low temperatures $P(V)$ is non-monotonic. Let us look at $P(V)$ at $T = 0.9\ T_C$ on which we put points, A, B, ..., G, by dots. The part (A–C) gives the gas phase which has a large volume in a given pressure, while the part (E–G) gives the liquid phase which has a small volume. In the part (C–E), $\left(\frac{\partial P}{\partial V}\right)_T > 0$, and this part is not allowed as a thermodynamic state. Thus, the system must jump at some pressure between a point in the branch of gas state (A–C) and a point in the liquid branch (E–G). The phase transition occurs when the Gibbs free energies of both points are the same. Thus, we need to find the pressure at which the Gibbs free energies are the same. In the figure, the points of the pressure are shown by F and B.

To determine the positions of F and B, we need to calculate the Gibbs free energy:

$$G(p) = G(p_A) + \int_{p_A}^{p} v(p')dp', \tag{2.70}$$

which is depicted in Fig. 2.17.

It should be noted that $v(p)$ is a non-monotonic function and the integral for the period (C–E) gives a negative contribution. The obtained $G(p)$ is also depicted in Fig. 2.17. In the figure, we find that the Gibbs free energy cross a point. At this point, the pressure is the same and also the Gibbs free energies are the same. In the equilibrium, the state with the minimum $G(p)$ is realized, and the state moves along A–B–F–G. Thus, the transition point is at this point, and we choose the points (B

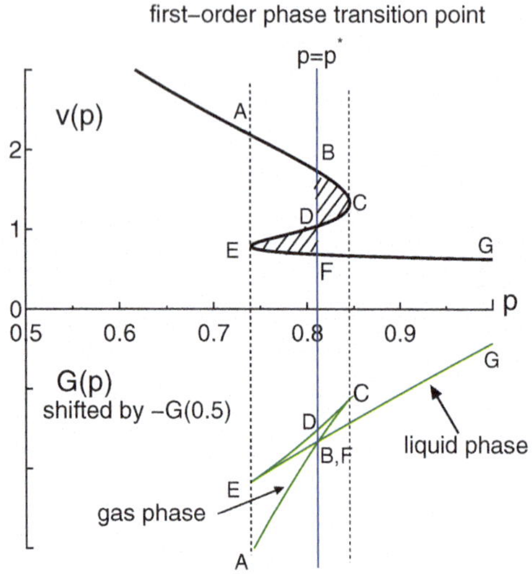

Fig. 2.17 Pressure dependence of the volume $v(p)$ and the Gibbs free energy $G(p)$

and F) for the first-order phase transition. The pressure of B and F gives the pressure p^* of the first-order phase transition. The part B–C and E–F are metastable states, which are called super-cooled gas and the over-heated liquid, respectively. The part C–E is thermodynamically unstable.

Because

$$\int_{B-C-E-D}^{p} v(p')dp' = 0, \tag{2.71}$$

the shaded areas BCD and DEF are the same. Therefore, we can determine P^* by choosing a pressure at which the areas are the same. This way to determine the first-order point is called 'Maxwell equal-area law'.

2.7 Statistical Treatments of the Gas-Liquid Phase Transition

2.7.1 Perturbational Approach

The partition function of the real gas (2.60) is given by

$$Z = \frac{1}{N!}\int d\boldsymbol{r}_1 \int d\boldsymbol{p}_1 \cdots \int d\boldsymbol{r}_N \int d\boldsymbol{p}_N e^{-\beta\mathcal{H}}. \tag{2.72}$$

In classical systems, the momentum part and interaction part are independent, and the part of momentum gives the partition function of the ideal gas $Z_{\rm ideal}$

$$Z_{\rm ideal} = \frac{V^N}{N!}\left(\frac{2\pi m k_{\rm B}T}{h^2}\right)^{3N/2}, \quad V = \int d\boldsymbol{r}. \tag{2.73}$$

For the real gas system, the integral over the interaction is not simply V^N but $V^N Q$:

$$Q = \frac{1}{V^N}\int d\boldsymbol{r}_1 \cdots \int d\boldsymbol{r}_N e^{-\beta\sum_{i,j}\phi(|\boldsymbol{r}_i - \boldsymbol{r}_j|)}, \tag{2.74}$$

which is called 'configuration integral'. Therefore, the partition function is given by

$$Z = Z_{\rm ideal} \times Q. \tag{2.75}$$

Then, the pressure P is given by

$$P = \frac{\partial k_{\rm B}T \ln Z}{\partial V} = N k_{\rm B}T\frac{1}{V} + \frac{\partial Q}{\partial V}. \tag{2.76}$$

It is difficult to calculate Q, and various approximate ways have been used. For example, Q is obtained in a series expansion of the density N/V [26]

$$\frac{PV}{Nk_{\rm B}T} = 1 + B_2(T)\left(\frac{N}{V}\right) + B_3(T)\left(\frac{N}{V}\right)^2 \ldots, \tag{2.77}$$

which is called Virial expansion, and the coefficients $B_j(T)$, $j = 2, 3, \ldots$ are called Virial coefficients. For example,

$$B_2(T) = -\frac{1}{2}\int d\boldsymbol{r}\left(e^{-\beta\phi(|\boldsymbol{r}|} - 1\right). \tag{2.78}$$

The expansion up to higher order can be done by making use of the cluster expansion method (Mayer expansion [27]).

2.7.2 Lattice-Gas Model Approach

Properties of phase transition can be estimated from the series expansion (2.77) by using some extrapolation methods. But here, we study the qualitative properties of phase transition in a simplified model called lattice-gas model, whose Hamiltonian is given by

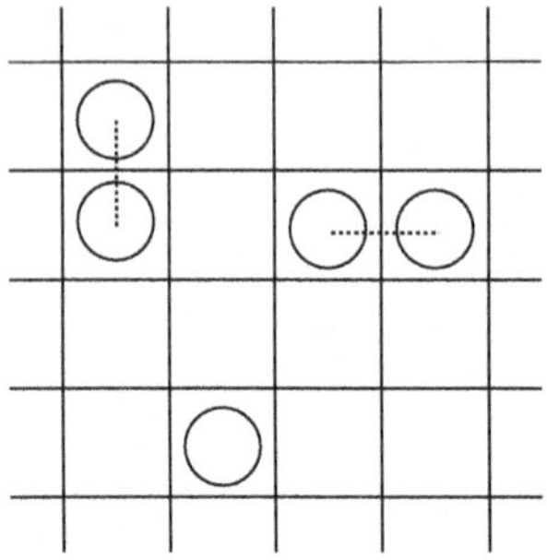

Fig. 2.18 Lattice gas model. The occupied cell is shown by a circle ($n_i = 1$), and the vacant cell denotes the variable $n_i = 0$. When the particles are close to each other, the interaction works as depicted by the dotted line

$$\mathcal{H}_{\text{lattice}} = -\phi_0 \sum_{<ij>} n_i n_j - \mu \sum_{i}^{N} n_j, \tag{2.79}$$

where n_i takes 0 or 1 at each cell of volume. The cell size is about that of a particle. The variable $n_i = 0$ or 1 corresponds to a vacant site and an occupied site, respectively (Fig. 2.18). The ingredients (1) and (2) of the real gas are taken into account in the model. Namely, the suppression of larger values of n_i, e.g., $n_i = 2, 3, \ldots$ corresponds to the effect of excluded volume, and the nearest neighbor interaction corresponds to the short-range interaction. μ is the chemical potential to control the density of the particle.

The variable n_i can be expressed by the Ising spin $\sigma_i (= \pm 1)$ as

$$n_i = \frac{1 + \sigma_i}{2} \quad \begin{cases} n_i = 1 \Leftrightarrow \sigma_i = 1 \\ n_i = 0 \Leftrightarrow \sigma_i = -1 \end{cases}, \tag{2.80}$$

and the Hamiltonian (2.79) is rewritten as

$$\mathcal{H}_{\text{lattice}} = -\frac{\phi_0}{4} \sum_{<ij>} \sigma_i \sigma_j - \left(\frac{z\phi_0}{4} + \frac{\mu}{2} \right) \sum_{i=1}^{N} \sigma_i. \tag{2.81}$$

Thus, the Hamiltonian (2.79) is equivalent to an Ising model with the parameters:

$$J_{\text{lattice}} = \frac{\phi_0}{4}, \quad H_{\text{lattice}} = \frac{z\phi}{4} + \frac{\mu}{2}. \tag{2.82}$$

In the lattice gas, the density of the particles $n = \langle n_i \rangle$ is related to the order parameter of the Ising model is $m = \langle \sigma_i \rangle$ as

$$n = \frac{1 + m}{2}. \tag{2.83}$$

Here it should be noted that the total number of spins N in the Ising model corresponds to the volume of the system in the lattice gas model and the number of the particles $N_{\rm particle}$ is given by $(N+M)/2$.

In this way, the gas–liquid transition is expressed by the ferromagnetic phase transition of the Ising model. As we see in the next subsection, at high temperatures, the density n is a monotonic function of μ and it becomes a non-monotonic function at low temperatures. The border gives the critical temperature. This scenario corresponds to the behavior of the magnetization as a function of the field.

2.7.3 Mean-Field Analysis for the Lattice Gas Model

The grand partition function for the model (2.81) is given by

$$\Xi \equiv \sum_{n_i=\pm 1} e^{-\beta\mathcal{H}} = \sum_{n_i=\pm 1} e^{\beta\phi_0 \sum_{<ij>} n_i n_j + \beta\mu \sum_i n_i}. \tag{2.84}$$

Here, we obtain the free energy by the mean-field method with the method ignoring the fluctuation. As in Appendix 8.1.2.1, we express the Hamiltonian by the deviation of the density from the average

$$\delta_i = n_i - n, \tag{2.85}$$

and ignore the second order of δ. Then, we have

$$-\beta\mathcal{H} \simeq V\frac{z}{2}\beta\phi_0 n^2 + zn\beta\phi_0 \sum_i \delta_i + \beta\mu \sum_i n_i, \tag{2.86}$$

which gives

$$\mathcal{H}_{\rm MF} = -(z\phi_0 n + \mu)\sum_i n_i + V\frac{z}{2}\phi_0 n^2. \tag{2.87}$$

Performing the sum over $n_i = 0, 1$, we obtain the grand partition function:

$$\Xi \equiv \prod_{i=1}^{V}\left(\sum_{n_i=0,1} e^{\beta(z\phi_0 n+\mu)\sum_i n_i - V\frac{z}{2}\beta\phi_0 n^2}\right) = e^{-\frac{V}{2}\beta z\phi_0 n^2}\left(1+e^{\beta(z\phi_0 n+\mu)}\right)^V. \tag{2.88}$$

Form (2.88), the average of n_i is given by

$$\langle N_{\rm particle}\rangle = \frac{\partial \log \Xi}{\partial \beta\mu} = V\frac{e^{\beta(z\phi_0 n+\mu)}}{1+e^{\beta(z\phi_0 n+\mu)}}. \tag{2.89}$$

Thus, we have the self-consistent equation:

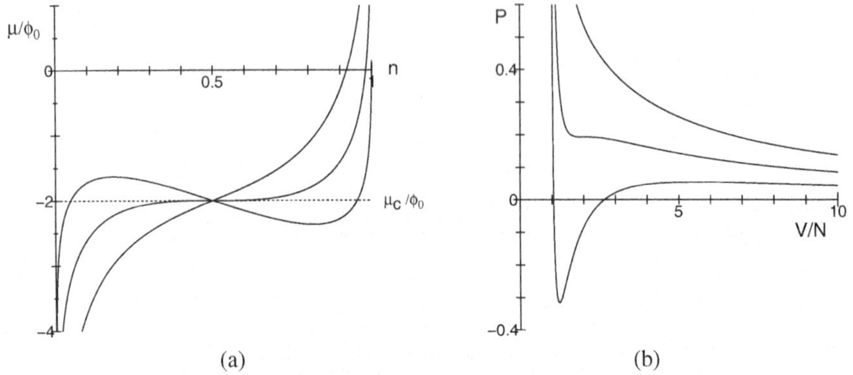

Fig. 2.19 **a** μ as a function of n at $T = 1.5\ T_C$ (monotonic), T_C, and $0.6\ T_C$ (non-monotonic, the part of negative slope is thermodynamically unstable), **b** PV phase diagram at $T = 1.5\ T_C$, T_C, and $0.6\ T_C$

$$n = \frac{\langle N_{\text{particle}} \rangle}{V} = \frac{e^{\beta(z\phi_0 n+\mu)}}{1+e^{\beta(z\phi_0 n+\mu)}} \to \mu = -z\phi_0 n + k_B T \log \frac{n}{1-n}. \qquad (2.90)$$

In Fig. 2.19a, we show n as a function of μ for $z = 4$ at $T = 1.5\ T_C$, T_C, $0.6\ T_C$, ($T_C = \frac{z}{4}\phi$). At high temperatures, n is a monotonic function of μ, while it is not monotonic below the critical temperature and three solutions appear corresponding to the metastable structure. For the given value of μ, the smallest solution of n gives the density of gas state, and the largest one is that of liquid, either of which is metastable. The middle one is a unstable solution.

By the self-consistent equation (2.90), the pressure P is given as a function of the volume V

$$\frac{1}{V}\ln \Xi = \frac{P}{k_B T} = -\frac{z\phi_0}{2k_B T}\left(\frac{N}{V}\right)^2 - \log\left(1-\frac{N}{V}\right) \qquad (2.91)$$

which is depicted in Fig. 2.19b, which is similar to the familiar van der Waals equations. At $T > T_C$, P is monotonic to V, while it is not at $T < T_C$.

Chapter 3
Escape Rate from the Metastable State

3.1 Introduction

In the mean-field picture, the metastable state survives until the so-called spinodal point. At this point, the metastable state collapses and changes to the equilibrium state discontinuously. In principle, the mean-field picture is valid when the system is uniform in the space and has a long-range interaction as explained in Appendix 8.1.

In the mean-field theory, the size of a system is assumed to be infinity, and the energy barrier is macroscopic, and thus the metastable state survives as long as a free energy barrier exists. Thus, approaching the spinodal point, there appears singular behavior of relaxation time, which is explained in Sect. 3.4. However, if the system size is finite, the barrier is finite even in systems with long-range interaction. Then, at finite temperatures, thermal fluctuation can cause a relaxation of the metastable state jumping over the potential barrier.

In contrast to the system with the long-range interaction, in short-range interacting systems, neighboring spins tend to align and form domains (or clusters) of the opposite magnetization as local excitations at finite temperatures. This cluster causes the breakout of the metastable state by a nucleation (Sect. 3.5).

Because critical nucleation is a local phenomenon, its excitation energy is finite, and thus the probability of the critical nucleation is finite regardless of the size of the system. Therefore, in the usual systems with short-range interactions, even before the spinodal point of the mean-field analysis, the metastable state collapses, i.e., metastable state relaxes in a finite time even in infinite systems. Thus, in principle, the metastable state is always destroyed by thermal fluctuation, the concept of metastability disappears, and the spinodal point cannot be defined.

Even in such a case, however, a sharp change in the relaxation time of the metastable state occurs. This point is called the dynamical spinodal point (Sect. 3.6). This point can be regarded as an effective spinodal point, and practical metastability may exist below this point. In magnets, a practical coercive field locates around this point. This idea is explained in Sect. 3.6.

S. Miyashita, *Collapse of Metastability*, Fundamental Theories of Physics 211,
https://doi.org/10.1007/978-981-19-6668-2_3

Even in a finite size short-range interacting system, is still so long that it is difficult to determine the first-order phase transition point. To overcome this difficulty, one can determine the transition point in numerical simulation by using the fact that the domain wall moves to expand the region of the stable phase if we prepare a configuration with an interface of both phases [28]. In this process, the difference in the free energies of phases causes the driving force.

3.2 Arrhenius Law

As for the decay of metastability, we generally use a picture of potential energy $U(x)$ like that depicted in Fig. 3.1.[1] In the figure, region A is a region for the metastable state. Let the initial state be in region A, and the the probability distribution of the particle be approximately given by a distribution $P_0(x)$ which is the thermal distribution in the potential $U_0(x)$ which is similar to $U(x)$ in region A but does not become small in region C.

$$P_0(x) \simeq \frac{1}{Z_0} e^{-\beta U_0(x)} \quad Z_0 = \int e^{-\beta U_0(x)} dx. \tag{3.1}$$

The probability distribution is localized around the bottom of the potential $U_0(x)$. But, on occasion that a particle comes near the top of the barrier of $U(x)$, i.e., region B, it may relax to the stable region C. Thus, the rate of relaxation r is expected for the population around the top of the barrier, i.e.,

$$r \sim e^{-\beta U_0(x_\mathrm{B})}, \tag{3.2}$$

which is the Arrhenius law of the escape rate.

This picture is often used to describe the dynamics of a state although a state of the system is not a particle. There, x is a parameter which specifies the state of the system, e.g., the magnetization of ferromagnetic systems, and $P(x)$ is regarded as the probability that the system is in the state being specified by x.

3.3 Kramers Method

Kramers studied the rate of escape from the metastable state in 1940 [1]. This is a pioneering study on the decay of metastable states. Here we give an outline of the study.

[1] The quantum dynamics of the wave function in this type of potential has been extensively studied as a scattering problem [30], which will be explained in PART III.

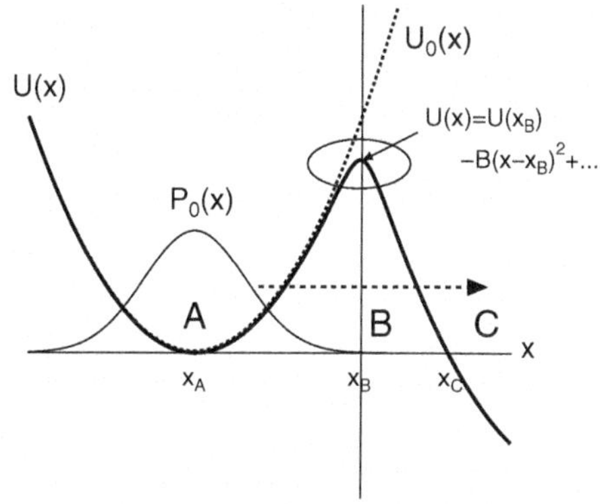

Fig. 3.1 Escape from the metastable state in region A over the free energy barrier in region B to the stable state in region C. The potential $U(x)$ is plotted by the bold line. $U_0(x)$ is an approximate potential in the metastable region (dotted curve). $P_0(x)$ is the initial distribution of probability trapped in the potential $U_0(x)$

The escape rate from the metastable state is expressed by a current of probability from the metastable state in region A to the stable state in region C through an energy barrier in region B. The time evolution of the probability distribution $P(x)$ follows the conservation law which is given in the form

$$\frac{\partial P(x)}{\partial m} = -\frac{\partial}{\partial x} J(x), \tag{3.3}$$

where $J(x)$ is the current of the probability. It is known that the dynamics of the probability distribution is given in the form (Fokker–Planck equation (see Appendix 8.2.1)):

$$\frac{\partial P(x)}{\partial x} = -\frac{\partial}{\partial m}\left[-h(x)P(x) + \varepsilon\frac{\partial}{\partial x}(D(x)P(x))\right], \tag{3.4}$$

where $h(x)$ is the drift term and it may be given by the derivative of the energy $U(x)$

$$h(x) = -\frac{\partial U(x)}{\partial x}, \tag{3.5}$$

and the second term represents diffusion. Here, we assume $D(x)$ is a constant and denote it by D. Then, the current is given in the form:

$$J(x) = -h(x)P(x) + \varepsilon D\frac{\partial}{\partial x}P(x) = e^{-U(x)/\varepsilon D}\frac{\partial}{\partial x}\left(e^{U(x)/\varepsilon D}\varepsilon D P(x)\right). \tag{3.6}$$

Thus, we have

$$J(x)e^{U(x)/\varepsilon D} = \frac{\partial}{\partial x}\left(e^{U(x)/\varepsilon D}\varepsilon D P(x)\right). \tag{3.7}$$

We assume the current is constant at all the m in a semi-steady state, i.e., $J(x) = J$, and integrate the above relation from A to C. Then, we have

$$J\int_{m_\mathrm{A}}^{m_\mathrm{C}} e^{U(x)/\varepsilon D}dx = \varepsilon D\, e^{U(x)/\varepsilon D}\varepsilon D P(x)\Big|_{m_\mathrm{A}}^{m_\mathrm{C}}. \tag{3.8}$$

In an early stage of escape, the system is almost in the metastable state A, and thus

$$e^{U(x_\mathrm{A})/\varepsilon D}P(x_\mathrm{A}) \simeq 1, \quad e^{U(x_\mathrm{C})/\varepsilon D}P(x_\mathrm{C}) \simeq 0. \tag{3.9}$$

Here, $U(x_\mathrm{A}) = U(x_\mathrm{C})$ and we set them to be 0 without loss of generality.

The integral in the right side of (3.8) is estimated by the saddle point method assuming that $U(m)$ is expressed near the maximum point B as

$$U(x) = U(x_\mathrm{B}) - B(x - x_\mathrm{C})^2, \quad \Delta U = U(x_\mathrm{B}). \tag{3.10}$$

Then,

$$\int_{m_\mathrm{A}}^{m_\mathrm{C}} e^{U(x)/\varepsilon D}dx \sim Ae^{\Delta U/\varepsilon D}, \tag{3.11}$$

where A is a perfector coming from the gauss-integration around m_C. In this way, the current is obtained as

$$J \propto \varepsilon A D e^{-\Delta U/\varepsilon D}. \tag{3.12}$$

Depending on the shape of $U(x)$, the perfector A changes, but the dominant contribution is the type of Arrhenius law.

3.4 Spinodal Singularity

As mentioned above, in the mean-field theory, the free energy gap is of order N and the lifetime of the metastable state is infinite in the thermodynamic limit. However, in finite size systems, the relaxation occurs by thermal fluctuation and the relaxation time is finite, and it shows peculiar size dependence near the spinodal point.

Here, we adopt the kinetic Ising model of the long-range interaction (Husimi–Temperley model (3.13)) which exhibits the mean-field type critical property (see Appendix 8.1.4):

$$\mathcal{H} = -\frac{Jz}{2N}\sum_{i=1}^{N}\sum_{i=1}^{N}\sigma_i\sigma_j - H\sum_{i=1}^{N}\sigma_i, \quad \sigma = \pm 1. \tag{3.13}$$

We study the dependence of the relaxation time on the system sizes around the spinodal point, using the Kramers type argument [23].

3.4.1 Master Equation for the Husimi–Temperley Model

For thermodynamic properties of the Ising model, we obtained the relation (2.4):

$$m = \tanh(\beta J z m + \beta H) \tag{3.14}$$

by the mean-field approximation.

To study the dynamics of the system, we need a concrete model of dynamics which exhibits this equation of state (3.14) as its stationary state. As explained in Appendix 8.1.4, the Husimi–Temperley model has this equation of state. Thus, we adopt this model and study its dynamics by a master equation which gives the time evolution of the probability of state i at time t. The master equation is generally given by the form (see Appendix 8.2)

$$\frac{\partial P(i,t)}{\partial t} = -\sum_j W(i \to j)P(i,t) + \sum_j W(j \to i)P(j,t), \tag{3.15}$$

where i and j are states of the system, and $W(i \to j)$ is the transition probability from the state i to j per unit time. We assume that the transition probability satisfies the detailed balance:

$$W(i \to j)P_{\rm eq}(i) = W(j \to i)P_{\rm eq}(j), \tag{3.16}$$

where $P_{\rm eq}(i)$ is the probability of the state i in equilibrium.

We study dynamics given by this master equation. The model is called the 'kinetic Ising model'. In the model, i denotes a spin state $(\sigma_1, \sigma_2, \ldots, \sigma_N)$, and the total number of states of the model (3.13) with N spins is 2^N, i.e., $i = 1, 2, \ldots, 2^N$. The time evolution of the probability distribution $(\{P(i)\})$ is given by (3.15). Here, we adopt the so-called single-slip-flip update, by which one of the spins, say σ_k, flips and the state changes as

$$i = (\sigma_1, \sigma_2, \ldots, \sigma_k, \ldots, \sigma_N) \to j = (\sigma_1, \sigma_2, \ldots, -\sigma_k, \ldots, \sigma_N)) \tag{3.17}$$

with a transition probability $W(i \to j)$.

Here, we adopt the Glauber model [29] in which $W(i \to j)$ is given as

$$W(i \to j) = \frac{1}{\tau_0}\frac{e^{-\beta E(j)}}{e^{-\beta E(i)} + e^{-\beta E(j)}}, \tag{3.18}$$

where τ_0 is a time scale of the process, e.g., $1/\tau_0$ is a frequency of contacts to the thermal bath per unit time.

The time evolution of (3.15) is given by a $2^N \times 2^N$ matrix. However, in the present model (3.13), dynamics of the system can be described by the motion of the magnetization:

$$M = \sum_i \sigma_i = -N, -N+2, \ldots, N, \tag{3.19}$$

instead of each spin configuration $(\sigma_1, \sigma_2, \ldots, \sigma_k, \ldots, \sigma_N)$ because of the mean-field nature of the model. Indeed, the Hamiltonian is given as a function of M as

$$\mathcal{H} = -\frac{Jz}{2N}M^2 - HM, \quad M = \sum_i \sigma_i = -N, -N+2, \ldots, N. \tag{3.20}$$

Here, it should be noted that the degeneracy of states for M is given by

$${}_N C_{N_+} = \frac{N!}{N_+!N_-!} = \frac{N!}{\left(\frac{N+M}{2}\right)!\left(\frac{N-M}{2}\right)!}, \tag{3.21}$$

where

$$N_+ = \frac{N+M}{2}, \quad N_- = \frac{N-M}{2} \tag{3.22}$$

are the number of up-spins and down-spins, respectively. In the description of dynamics using M, the number of states is reduced from 2^N to $N+1$. This reduction is good only for models of the mean-field type where all the spins interact with each other uniformly.

Here, we follow the analysis given in the paper [23]. The master equation for the probability as a function of M is given by

$$\frac{\partial P(M,t)}{\partial t} = -W(M \to M+2)P(M,t) - W(M \to M-2)P(M,t)$$

$$+ W(M+2 \to M)P(M+2,t) + W(M-2 \to M)P(M-2,t). \tag{3.23}$$

It should be noted that in the above equation, we use the same characters W as those in (3.15) for the transition probability but they are different. The first term corresponds to the process $M \to M-2$ flipping one of the $+$ spins and the second term to the process $M \to M+2$ flipping one of the $-$ spins in the state of M. And the third term corresponds to the process $M+2 \to M$ flipping one of the $+$ spins, and the last term the process $M-2 \to M$ flipping one of the $-$ spins. The transition probability of the process $M \to M-2$ is

$$W(M \to M+2) = N_+ \frac{1}{\tau_0} \frac{e^{\beta \frac{Jz}{2N}(M+2)^2 + H(M+2)}}{e^{\beta \frac{Jz}{2N}(M+2)^2 - H(M+2)} + e^{\beta \frac{Jz}{2N}M^2 - HM}}$$

$$= \frac{N+M}{2\tau_0} \frac{e^{\beta \frac{Jz}{2N}(2M+4) + 2H)}}{e^{\beta \frac{Jz}{2N}(2M+4) + 2H)} + 1} = \frac{N+M}{2} \frac{e^{\beta(J(M-1)/N + H)}}{2\cosh(\beta(J(M-1)/N + H))}. \tag{3.24}$$

With similar calculations for other transition probabilities, the master equation is explicitly given as

$$\tau_0 \frac{\partial P(M,t)}{\partial t} =$$

$$-\frac{N+M}{2}\frac{e^{\beta(Jz(M-1)/N+H)}}{2\cosh(\beta(Jz(M-1)/N+H))}P(M,t)$$

$$-\frac{N-M}{2}\frac{e^{\beta(Jz(M+1)/N+H)}}{2\cosh(\beta(Jz(M+1)/N+H))}P(M,t)$$

$$+\frac{N+M+2}{2}\frac{e^{\beta(Jz(M+1)/N+H)}}{2\cosh(\beta(Jz(M+1)/N+H))}P(M+2,t)$$

$$+\frac{N-M+2}{2}\frac{e^{\beta(Jz(M-1)/N+H)}}{2\cosh(\beta(Jz(M-1)/N+H))}P(M-2,t). \tag{3.25}$$

For large N, we set

$$m = \frac{M}{N} \tag{3.26}$$

and regard m as a continuous variable, and expand (3.25) in a series of

$$\varepsilon = \frac{1}{N}. \tag{3.27}$$

Then, we have the following Fokker–Planck equation

$$\frac{\partial P(m)}{\partial t} = \frac{\partial}{\partial m}h(m)P(m) + \varepsilon\frac{\partial^2}{\partial m^2}D(m)P(m), \tag{3.28}$$

where the drift velocity is

$$h(m) = m - \tanh(\beta(Jzm + H) + \varepsilon\frac{\beta Jzm}{\cosh(\beta(Jzm + H))} \tag{3.29}$$

and the diffusion coefficient is

$$D(m) = 1 - m\tanh(\beta(Jzm + H)). \tag{3.30}$$

3.4.1.1 Critical Behavior Around the Spinodal Point

Near the spinodal point, we set

$$x = m - m_{\mathrm{SP}}, \quad y = \beta(H - H_{\mathrm{SP}}) \equiv h - h_{\mathrm{SP}} \tag{3.31}$$

with H_{SP} (2.12) and m_{SP} (2.13). Up to the second order of x and y, $h(m)$ is

$$h(m) = -\frac{y}{\beta Jz} - \eta(\beta Jzx^2 + 2xy), \tag{3.32}$$

where

$$\eta = |m_{\mathrm{SP}}| = \sqrt{1 - \frac{1}{\beta Jz}}, \tag{3.33}$$

and $D(m)$ is given by

$$D(m) = \frac{1}{\beta Jz}. \tag{3.34}$$

Now we study the relaxation time from the spinodal point $m(0) = m_{\mathrm{SP}}$, that is $x(0) = 0$. When x increases and becomes to be order $\sqrt{y}$, the term $\eta\beta Jzx^2$ becomes relevant. In this case, the term ηxy is small, and thus to study the relaxation from $x = 0$ we may discard this term, and then

$$h(m) = -\frac{y}{\beta Jz} - \eta\beta Jzx^2. \tag{3.35}$$

The corresponding Fokker–Planck equation is given by

$$\frac{\partial P(x,t)}{\partial t} = \left\{ -\frac{\partial}{\partial x}\left(\frac{y}{\beta Jz} + \eta\beta Jzx^2\right) + \frac{\varepsilon}{\beta Jz}\frac{\partial^2}{\partial x^2} \right\} P(x,t). \tag{3.36}$$

Introducing scaled parameters

$$\xi = x|y|^{-1/2}, \quad \Lambda = \varepsilon^{-2/3}y, \quad s = \varepsilon^{1/3}t, \tag{3.37}$$

we have the scaled Fokker–Planck equation

$$\frac{\partial P(\xi,t)}{\partial t} = \varepsilon^{1/3}\left\{ -|\Lambda|^{1/2}\frac{\partial}{\partial \xi}\left(\pm\frac{y}{\beta Jz} + \eta\beta Jz\xi^2\right) + \frac{1}{\beta Jz|\Lambda|}\frac{\partial^2}{\partial \xi^2} \right\} P(\xi,t). \tag{3.38}$$

For $\Lambda > 0$, the upper sign is applied, while for $\Lambda < 0$ the lower sign is applied. Because Eq. (3.38) depends only on Λ, is expected to have the scaling form [23]

$$\tau = \varepsilon^{-1/3}f(\Lambda) = N^{1/3}f(N^{2/3}(h - h_{\mathrm{SP}})), \tag{3.39}$$

where f is a scaling function.

The relaxation time for the unstable case, i.e., $y > 0$, is finite even in the case $N = \infty$. There, we put $\varepsilon = 0$ in (3.38), and the Fokker–Planck equation is

$$\frac{\partial P(\xi,t)}{\partial t} = -y^{1/2}\frac{\partial}{\partial \xi}\left(\frac{1}{\beta J z} + \eta\beta J z\xi^2\right)P(\xi,t). \tag{3.40}$$

Therefore, the relaxation time should be scaled like

$$\tau \sim y^{-1/2} \sim (h - h_{\rm SP})^{-1/2}, \tag{3.41}$$

which diverges when $h \to h_{\rm SP}$.

At the spinodal point, the relaxation time diverges as N increases [31]:

$$\tau \sim N^{1/3}. \tag{3.42}$$

The relaxation time for the metastable case, i.e., $y < 0$, is estimated by the Kramers type argument (Sect. 3.3) [1, 32]. When $y < 0$, we rewrite (3.38) as

$$\frac{\partial P(\xi,t)}{\partial t} = \left\{-|\Lambda|^{1/2}\frac{\partial}{\partial \xi}\left(\pm\alpha + \gamma\xi^2\right) + \frac{\delta}{|\Lambda|}\frac{\partial^2}{\partial \xi^2}\right\}P(\xi,t), \tag{3.43}$$

where

$$\alpha = \frac{1}{\beta J z}, \quad \gamma = \eta\beta J z, \quad \delta = \frac{1}{\beta J z}. \tag{3.44}$$

Introducing an explicit form of U

$$U(\xi) = \alpha\xi - \frac{1}{3}\gamma\xi^3, \tag{3.45}$$

it is written as

$$\frac{\partial P(\xi,t)}{\partial t} = \varepsilon^{1/3}\frac{\partial}{\partial \xi}\left(\frac{\delta}{|\Lambda|}e^{-|\Lambda|^{3/2}U(\xi)/\delta}\frac{\partial}{\partial \xi}e^{|\Lambda|^{3/2}U(\xi)/\delta}P(\xi,t)\right). \tag{3.46}$$

The current J is given by

$$J/\varepsilon^{1/3} = -\frac{\delta}{|\Lambda|}e^{-|\Lambda|^{3/2}U(\xi)/\delta}\frac{\partial}{\partial \xi}e^{|\Lambda|^{3/2}U(\xi)/\delta}P(\xi,s) \tag{3.47}$$

and the stationary current is

$$J/\varepsilon^{1/3} = -\frac{\delta}{|\Lambda|}\frac{e^{|\Lambda|^{3/2}U(\xi)/\delta}P(\xi,s)\Big|_{\rm A}^{\rm B}}{\int_{\rm A}^{\rm B} e^{|\Lambda|^{3/2}U(\xi)/\delta}d\xi}. \tag{3.48}$$

Estimating the integral by the saddle point method at the top of the barrier C:

$$\int_{A}^{B} e^{-|\Lambda|^{3/2}U(\xi)/\delta} d\xi \simeq \sqrt{\frac{-2\pi\delta}{|\Lambda|^{3/2}U''(C)}} e^{|\Lambda|^{3/2}U(C)/\delta} \tag{3.49}$$

and for early stage, setting $P(B) = 0$, we have

$$J = P(A)\varepsilon^{1/3}\sqrt{\frac{-U''(C)\delta}{2\pi|\Lambda|^{1/2}}} e^{-|\Lambda|^{3/2}(U(C)-U(A))/\delta}. \tag{3.50}$$

Here, $P(\xi)$ is given by

$$P(\xi) = \frac{\exp\left[-\frac{|\Lambda|^{3/2}}{\delta}\left\{U(A) + \frac{1}{2}U''(A)(\xi - A)^2\right\}\right]}{Z}, \tag{3.51}$$

where

$$Z = \int_{-\infty}^{\infty} \exp\left[-\frac{|\Lambda|^{3/2}}{\delta}\left\{U(A) + \frac{1}{2}U''(A)(\xi - A)^2\right\}\right] d\xi$$

$$= e^{-|\Lambda|^{3/2}U(A)/\delta}\sqrt{\frac{2\pi\delta}{|\Lambda|^{3/2}U''(A)}}. \tag{3.52}$$

Thus,

$$P(A) = \frac{\exp\left[-\frac{|\Lambda|^{3/2}}{\delta}\{U(A)\}\right]}{Z} = \sqrt{\frac{|\Lambda|^{3/2}U''(A)}{2\pi\delta}}. \tag{3.53}$$

The relaxation time τ is

$$\tau \equiv \frac{1}{J} \sim N^{1/3}2\pi|U''(A)U''(C)|^{-1/2}|\Lambda|^{-1/2}\exp\left[\frac{|\Lambda|^{3/2}}{\delta}(U(C) - U(A))\right]. \tag{3.54}$$

From (3.45),

$$A = -C = \frac{\alpha}{\gamma}, \quad U(C) - U(A) = \frac{4\alpha}{3}\sqrt{\frac{\alpha}{\gamma}}, \quad U''(A) = -U''(C) = 2\sqrt{\alpha\gamma}. \tag{3.55}$$

Substituting these, the relaxation time is

$$\tau \sim N^{1/3} \frac{\pi}{2\sqrt{\eta|\Lambda|}} \exp\left(\frac{4}{3\beta J z \sqrt{\eta}} |\Lambda|^{3/2} \right). \tag{3.56}$$

It should be noted that

$$\frac{4}{3\beta J z \sqrt{\eta}} |\Lambda|^{3/2} = \beta[F(C) - F(A)] = \beta \Delta F, \tag{3.57}$$

that is, ΔF is the difference between the free energies at the maximum point and at the metastable point.

Thus, we have

$$\tau \sim N^{1/3} \frac{\pi}{2\sqrt{\eta|\Lambda|}} \exp(\beta \Delta F), \tag{3.58}$$

where ΔF is proportional to N. This gives the size dependence of the relaxation time at the spinodal point in the mean-field model.

3.4.1.2 Method Using WKB Approximation

The escape rate is also obtained by using the WKB approximation for a kind of Schrödinger equation which is transformed from the Fokker–Planck equation [32]. We assume that the Fokker–Planck equation

$$\frac{\partial P}{\partial t} = \frac{\partial}{\partial x}\left[\frac{\partial U}{\partial x} P\right] + \varepsilon D \frac{\partial^2}{\partial x^2} P = \frac{\partial}{\partial x}\left[e^{-U/\varepsilon D} \frac{\partial}{\partial x}\left(e^{U/\varepsilon D} \varepsilon D P\right)\right] \tag{3.59}$$

has a stationary state

$$P_{\rm eq}(x) = \frac{1}{Z} e^{-U(x)/\varepsilon D}. \tag{3.60}$$

Using the transformation

$$P(\xi) = e^{-\frac{1}{2}U(\xi)/\varepsilon D} Q(\xi), \tag{3.61}$$

the Fokker–Planck equation is transformed to a self-adjoint form

$$\frac{\partial Q}{\partial t} = \left(-\frac{1}{4\varepsilon D}\left(\frac{\partial U}{\partial x}\right)^2 + \frac{1}{2}\varepsilon D \frac{\partial^2 U}{\partial x^2}\right) Q(x) + \varepsilon D \frac{\partial^2}{\partial x^2} Q(x). \tag{3.62}$$

The Schrödinger equation of a particle with the mass εD

$$i\hbar \frac{\partial \Psi}{\partial t} = \left(-\frac{\hbar^2}{\varepsilon D} \frac{\partial^2}{\partial x^2} + V(x)\right) \Psi \tag{3.63}$$

is transformed to

$$\frac{\partial \Psi}{\partial \tau} = \left(\frac{\hbar^2}{\varepsilon D} \frac{\partial^2}{\partial x^2} - V(x) \right) \Psi, \tag{3.64}$$

for the imaginary time $\tau \equiv it/\hbar$. Thus, effective potential in (3.62) is

$$V(x) = \frac{1}{4\varepsilon D} \left(\frac{\partial U}{\partial x} \right)^2 - \frac{1}{2} \varepsilon D \frac{\partial^2 U}{\partial x^2}. \tag{3.65}$$

With this potential, we can estimate the relaxation time from the second lowest eigenvalue. By the WKB approximation (Appendix 8.6), the relaxation time is estimated as [23]

$$\tau \sim 2N^{1/3} \sqrt{\pi} |U''(A)U''(C)|^{-1/2} |\Lambda|^{-1/2} e^{\beta \Delta F}. \tag{3.66}$$

3.5 Nucleation in Model Short-Range Interaction

We have studied the spinodal point as an end point of metastability in the mean-field analysis in the previous chapter. However, as well-known, for systems with short-range systems, the metastable state is broken by local breakdown, i.e., a nucleation of stable phase.

In the metastable ordered state, while small anti-magnetized clusters tend to vanish due to the surface tension, and large clusters tend to expand due to the Zeeman energy which favors the opposite magnetization. Thus, a certain threshold of size exists. Beyond the threshold, the gain of the Zeeman energy becomes larger than that of loss due to the surface. The cluster of the threshold size is called 'critical nucleus'. Once the critical nucleus appears by thermal fluctuation, the metastable state is broken.

We will call the occurrence of critical nucleus 'critical nucleation', or simply 'nucleation'. The nucleation occurs with a finite energy barrier, and thus even in an infinite size system, the metastable state decays to the stable state in a finite time at finite temperatures.

Let us consider a nucleation of a radius r. For concreteness, we consider the system in three dimensions. In this case, the loss of energy due to the surface is $\Delta E_J (\simeq 4\pi r^2 J)$, and the gain of the Zeeman energy is $\Delta E_H (= \frac{4}{3}\pi r^3 H)$. Thus, the excess energy due to the nucleation is

$$\Delta E = -\Delta E_H + \Delta E_J = -\frac{4}{3}\pi r^3 H + 4\pi r^2 J \tag{3.67}$$

which is depicted in Fig. 3.2. The maximum point gives the critical radius r_C and height is the energy gap ΔE_C. They are obtained by the relation

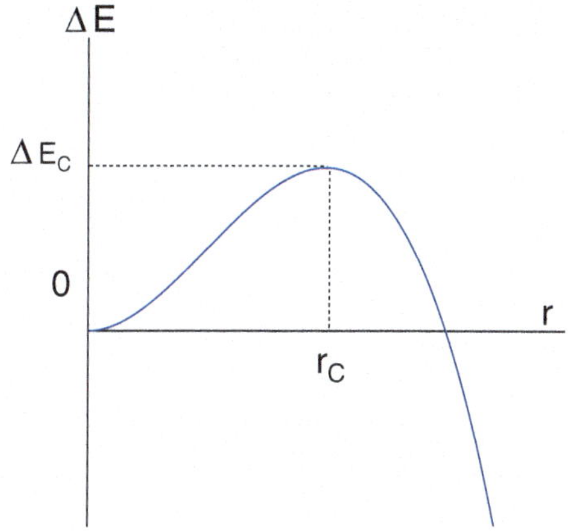

Fig. 3.2 Dependence of the excess energy ΔE due to nucleation of the radius r. The energy and radius of the critical nucleation are ΔE_c and r_c, respectively

$$\frac{d\Delta E}{dr} = -4\pi r^2 H + 8\pi r J = 0 \tag{3.68}$$

as

$$r_C = \frac{2J}{H}, \quad \Delta E_C = \frac{4\pi J}{3}\left(\frac{2J}{H}\right)^2. \tag{3.69}$$

For $r < r_C$, the nucleation tends to shrink and the system remains in the metastable state, while for $r > r_C$, the nucleation expands and brings the system to the stable (equilibrium) state. The thermal fluctuation can create nucleation of r_C jumping up to the barrier. At a finite temperature T, the probability to have the nucleation is

$$p_{\text{nucleation}} \propto e^{-\beta \Delta E_C}, \quad \beta = \frac{1}{k_B T}, \tag{3.70}$$

and the relaxation time of the metastable state is given by

$$\tau = p_{\text{nucleation}}^{-1} \tag{3.71}$$

which is always finite for systems at $T > 0$.

Therefore, in principle, the relaxation time is always finite and the concept of metastability disappears, and the metastability cannot be defined in contrast to the case of the mean-field models. However, in nature, even in systems with short-range interactions, metastable-like phenomena appear. For those cases, we have to investigate metastability as a problem of slow relaxation time.

3.6 Dynamical Spinodal Point

As we discussed in the previous subsection, the relaxation time of a metastable state is always finite in systems with short-range interaction. However, even in such systems, we find apparent metastable phenomena. For example, in the gas–liquid transition we really find the overheated liquid and the supercooled vapor, and also, we find coercivity in magnets. In Monte Carlo simulations of the Ising model below the critical temperature, we find the magnetization survives in an antiparallel field when the field is weak. In these systems, the relaxation time largely changes with the amplitude of the field, although the dependence is not singular.

Let us consider the relaxation of the metastable state in the Ising model with nearest-neighbor interactions. For a given size of systems, there are two characteristic regions of the field as schematically depicted in Fig. 3.3 [33, 34].

One of them is the multi-nucleation region at a rather large field. There, many nucleations occur before the magnetization nucleated in the first nucleation propagates to the border (Fig. 3.3 (left)). This process is called the Avrami process, where the relaxation times do not distribute largely because of the self-averaging property, and the process is called ‘deterministic’.

The other is the single-nucleation region at a small field. In this case, the probability of a nucleation is very small, and the magnetization nucleated in the first nucleation propagates to the border of the system before another nucleation occurs (Fig. 3.3 (right)). In this case, the relaxation occurs by the rate of nucleation. Because the nucleation occurs with the probability of an exponential type (3.70), relaxation times distribute largely ($\sim 1/p_{\rm nucleation}$), and the process is called ‘stochastic’. Between these regions, there is a point where the relaxation time increases very rapidly which is called the ‘dynamical spinodal point’, below which the apparent metastable state exists with a very long relaxation time [33, 34].

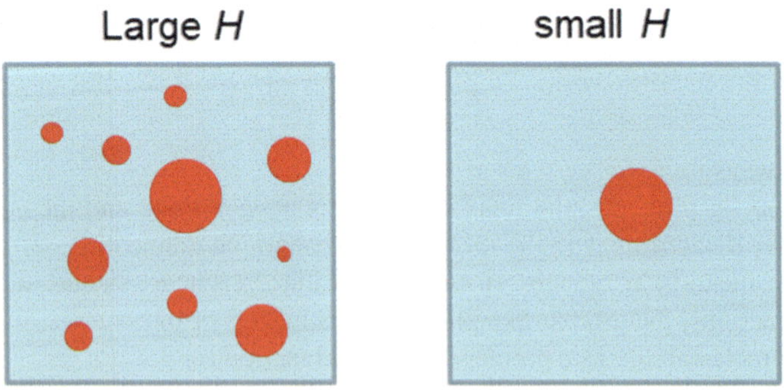

Fig. 3.3 The multi-nucleation region (left) and the single-nucleation region (right)

3.7 Survival Probability of a Metastable State

We consider the relaxation process in the stochastic region, i.e., in an apparent metastable state. In this subsection, we simply express $p_{\mathrm{nucleation}}$ by p, and give it by

$$p = ae^{-\beta E_{\mathrm{B}}(H)}. \tag{3.72}$$

Here, a is a perfector which represents the number of contacts with the thermal bath in unit time, and as a typical value it is often set to be

$$a = 10^{11}. \tag{3.73}$$

$E_{\mathrm{B}}(H)$ is the energy (or free energy) gap in a magnetic field H. The probability that no nucleation occurs in a small time interval $\Delta t (\ll 1)$ is given by

$$p_{\text{no nucleation}} = 1 - ae^{-\beta E_{\mathrm{B}}(H)}\Delta t \simeq \exp\left(-ae^{-\beta E_{\mathrm{B}}(H)\Delta t}\right). \tag{3.74}$$

3.7.1 Néel–Arrhenius Process

Because the nucleations occur stochastically, times of reversal of magnetization distribute. This effect was studied by Néel [35] and Brown [36]. Under a sweeping field, the distribution depends on the sweeping velocity [37].

Now, we consider the dynamics under a sweeping field $H(t)$ from a large negative value to a positive value $H(t)$ and obtain the survival probability $P(t)$, i.e., the probability that the metastable state survives until the time t. Let the energy barrier at a given value of field H be $E_{\mathrm{B}}(H(t))$ and discretize the time by an interval Δt. Then, the survival probability is given by

$$P(t = n\Delta t) = \prod_n \left[1 - ae^{-\beta E_{\mathrm{B}}(H(n\Delta t))}\Delta t\right] = \exp\left[-a \int_{-\infty}^{t} e^{-\beta E_{\mathrm{B}}(H(t'))} dt'\right]. \tag{3.75}$$

Here, we consider a simple case, where the field is swept linearly,

$$H(t) = vt, \tag{3.76}$$

and then the probability is given as a function of H:

$$P(H) = \exp\left[-a \int_{H(-\infty)=-\infty}^{H(t)} e^{-\beta E_{\mathrm{B}}(h)} \frac{dt}{dh} dh\right] = \exp\left[-\frac{a}{v} \int_{-\infty}^{H} e^{-\beta E_{\mathrm{B}}(h)} dh\right]. \tag{3.77}$$

This function is easily calculated numerically for any form of $E_B(H)$. If we give some explicit form, we may analytically integrate it. For example, let us consider the following form:

$$E_B(H) = E_0\left(1 - \frac{H}{H_0}\right)^n, \tag{3.78}$$

which is often used in the analyses of magnets. Here, we take the value of n as

$$n = 1, \tag{3.79}$$

which is often found in studies of magnets. Then, the explicit form of $P(H)$ is given as

$$P(H) = \exp\left[-\frac{aH_0}{v\beta E_0} e^{-\beta E_0\left(1-\frac{H}{H_0}\right)}\right]. \tag{3.80}$$

This form is sometimes called the 'Néel–Arrhenius function'. We depict the form for $v = 0.0001, 0.001$ and 0.01 in Fig. 3.4. When the sweep velocity is large, the relaxations occur at a large field.

The rate of relaxation at H is given by

$$\frac{dP(H)}{dH} = -\frac{a}{v} e^{-\beta E_0\left(1-\frac{H}{H_0}\right)} \exp\left[-\frac{aH_0}{v\beta E_0} e^{-\beta E_0\left(1-\frac{H}{H_0}\right)}\right], \tag{3.81}$$

which is also plotted.

If we sweep the field forward and backward, we have a hysteresis as depicted in Fig. 3.5, where the magnetization is given by

$$M(H) = 1 - 2P(H). \tag{3.82}$$

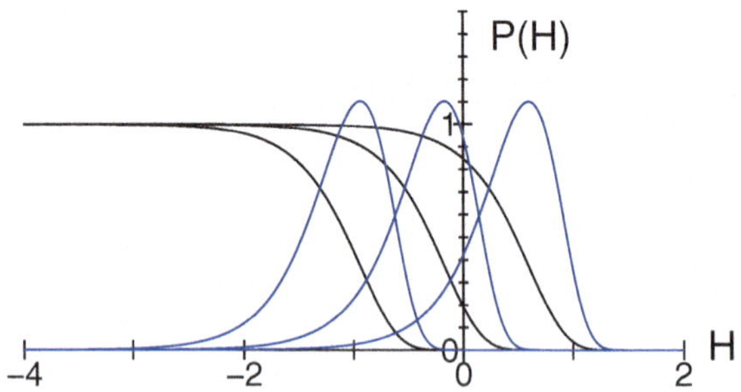

Fig. 3.4 Change of the survival probability $P(H)$ (3.80) in a swept field $H(t) = vt$ for $v = 0.0001, 0.001$ and 0.01 from left to right. Here, $\beta = 3$ and $E_0 = H_0 = 1$. The corresponding derivatives (3.81) are also plotted. The formula is valid for $H > H_0$

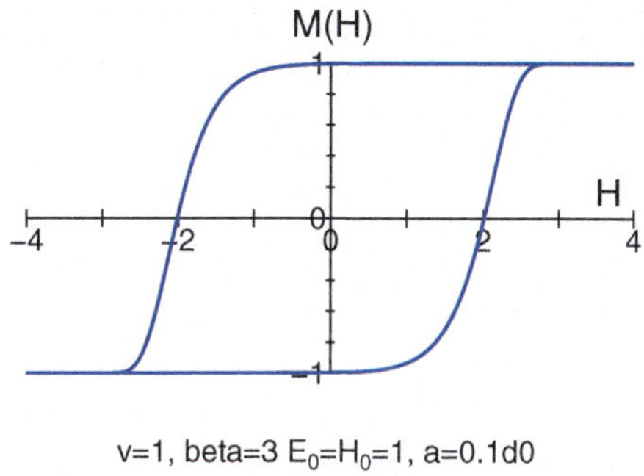

Fig. 3.5 Neel–Arrhenius hysteresis

This hysteresis is similar to that obtained in the mean-field model (Fig. 2.3a), but it should be noted that they are conceptually different.

In each observation of the sweeping process, the magnetization shows a jump, and the distribution is obtained by the repetition of such observations. Experimentally, the distribution of the times at which the magnetization of a single small magnet jumps was observed by Wernsdorfer et al. by using the microSQUID technique [38], and also the distribution for an ensemble of grains of $Ne_2Fe_{14}B$ magnets by Yomogita et al. [39].

Chapter 4
Spatial Pattern During the Transition

4.1 Dynamics Associated with the First-Order Phase Transition

There are two types of processes of formation of the ordered state, i.e., the field-quench process and the temperature-quench process. The former is a process in which the field changes the sign below the critical temperature. In this case, the metastable state is destroyed by nucleations, which were studied in the previous chapter and schematically depicted in Fig. 3.3. The latter is a process in which the temperature is quenched from the paramagnetic phase to the ordered phase beyond the critical temperature. At high temperatures, two components $\sigma_i = \pm 1$ are randomly mixed. When the temperature goes down below the critical temperature, the spins tend to align locally, and domains of up-spins and those of down-spins are formed, and the pattern of domains evolves over time. In this chapter, we mainly study this temperature-quench case [40].

4.2 Dynamics After the Temperature Quenching

The Ising model can be a model for the magnetic systems and also binary alloys. In the former, $\sigma_i = \pm 1$ represents spin up down, and the sum

$$M = \sum_{i=1}^{N} \sigma_i \tag{4.1}$$

can change by a flip of each spin. In this case, the model is called a non-conserved system. On the other hand, in the latter, M is conserved because the number of atoms is fixed and the ordering process shows the segregation of two components. In this

S. Miyashita, *Collapse of Metastability*, Fundamental Theories of Physics 211,
https://doi.org/10.1007/978-981-19-6668-2_4

case, a change of the configuration occurs by an exchange of $\pm$ spins, and the model is called a conserved system.

4.2.1 Non-conserved System: k^2t Scaling

In the case of non-conserved systems, each spin may change the sign locally, i.e., $\sigma_i = 1 \leftrightarrow -1$, and $m = \sum_i \sigma_i/N$ can change, and the dynamics of m is often given by the Van Hove type equation (8.13)

$$\frac{dm}{dt} = -\frac{1}{\tau_0}\frac{df(m)}{dm} + \text{noise term},$$

where $f(m)$ is the free energy of a function of m. This equation means that $m(t)$ is moved by the thermodynamic force $df(m)/dm$ without inertia, i.e., an overdamped motion. In inhomogeneous systems, the magnetization is a function of the position $\boldsymbol{r}$, and we use the magnetization profile $m(\boldsymbol{r})$. Then, the free energy is extended to a form including the stiffness term:

$$F(m(\boldsymbol{r})) = \int d\boldsymbol{r}\left[\frac{1}{2}K(\nabla m(\boldsymbol{r}))^2 + f(m(\boldsymbol{r}))\right], \tag{4.2}$$

where $f(m)$ is the local free energy as a function of magnetization for which we take the same form as that in the uniform system, and K is the stiffness constant which requires a smooth variation of $m(\boldsymbol{r})$. Then, we have the equation of motion of $m(\boldsymbol{r})$ by the functional derivative as

$$\frac{dm(\boldsymbol{r})}{dt} = -\frac{1}{\tau_0}\left[-K\nabla^2 m(\boldsymbol{r}) + \frac{d}{dm}f(m(\boldsymbol{r}))\right]. \tag{4.3}$$

This equation is called the time-dependent Ginzburg–Landau (TDGL) equation.

When a random state is quenched to a temperature below the critical point, first locally ordered clusters with up- or down-magnetization are formed from place to place. This early process is expressed by the dynamics of uniform magnetization (4.2.1). Relaxation process from the unstable point ($m = 0$) to the ordered state ($m = \pm m_s$) has been studied by Suzuki [41], where the so-called Suzuki-scaling process takes place. The process for the formation of a local ordered cluster is given by the Van Hove equation for the system with $f(m) = -\gamma m^2/2 + gm^4/4$:

$$\frac{dm}{dt} = \gamma m - gm^3 + \eta(t), \quad \gamma > 0, g > 0. \tag{4.4}$$

Here, a random noise $\eta(t)$

$$\langle \eta(t) \rangle = 0, \quad \langle \eta(t)\eta(t') \rangle = 2\varepsilon\delta(t - t') \tag{4.5}$$

is included because the initial state $m = 0$ is just on the unstable point, i.e., $dm/dt = df(m)/dm = 0$ and thus the state $m = 0$ does not move without noise.

The formal solution is given by

$$m(t) = e^{\gamma t} \int_0^t e^{-\gamma t'} \left(\eta(t') - g m^3(t') \right) dt'. \tag{4.6}$$

We solve this in perturbation series of g.

$$m(t) = e^{\gamma t} \int_0^t e^{-\gamma t'} \eta(t') dt'$$

$$- g e^{\gamma t} \int_0^t e^{-2\gamma t'} \int_0^{t'} dt_1 \int_0^{t'} dt_2 \int_0^{t'} dt_3 e^{-\gamma(t_1+t_2+t_3)} \eta(t_1)\eta(t_2)\eta(t_3) + \cdots . \tag{4.7}$$

Averaging over $\eta(t)$, odd moments disappear, e.g., $\langle m(t) \rangle = 0$. Using (4.5), the second moment $\langle m(t)^2 \rangle$ is expressed as

$$\langle m(t)^2 \rangle = \frac{\varepsilon}{\gamma} \left(e^{2\gamma t} - 1 \right) - 3\frac{\varepsilon^2}{\gamma^3} \left(e^{4\gamma t} - 4\gamma t e^{2\gamma t} - 1 \right) \ldots . \tag{4.8}$$

Introducing a scale variable τ:

$$\tau = \frac{g\varepsilon}{\gamma^2} e^{2\gamma t}, \tag{4.9}$$

and taking the limit $t \to \infty$

$$\langle m(t)^2 \rangle = \langle m(t)^2 \rangle_{\mathrm{st}} \left(\tau - 3\tau^2 + 15\tau^3 + \cdots \right), \tag{4.10}$$

where $\langle m(t)^2 \rangle_{\mathrm{st}} = \frac{\varepsilon}{\gamma}$. The other terms are also estimated, and they are given as

$$\langle m(t)^2 \rangle = \langle m(t)^2 \rangle_{\mathrm{st}} \sum_{n=1}^{\infty} (-1)^{n-1} (2n-1)!! \tau^n. \tag{4.11}$$

This is an asymptotic series, but it can be expressed by using the Borel sum:

$$\langle m(t)^2 \rangle = \langle m(t)^2 \rangle_{\mathrm{st}} \frac{1}{\sqrt{2\pi}} \int_{-\infty}^{\infty} e^{-\xi^2/2} \frac{\xi^2 \tau}{1 + \xi^2 \tau} d\xi. \tag{4.12}$$

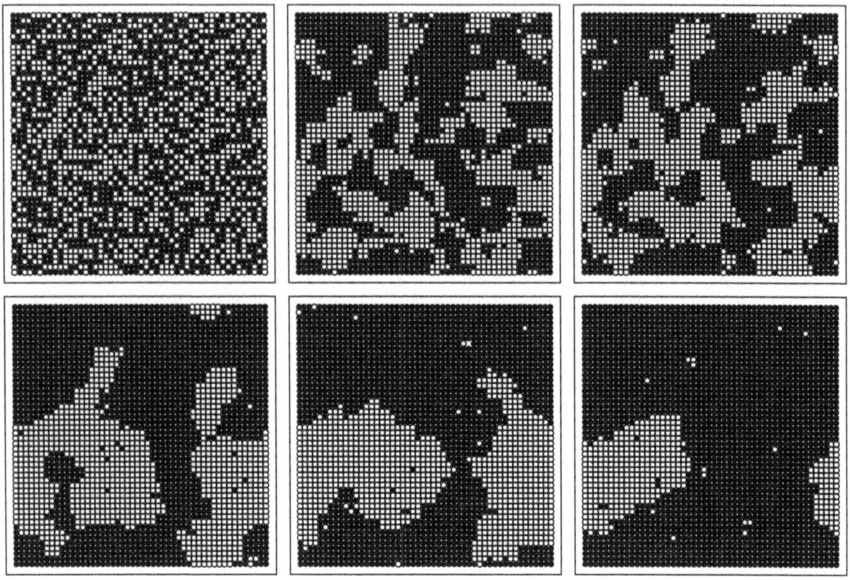

Fig. 4.1 Development of configurations after a quench of temperature below the critical point in the non-conserved system. $t = 0, 5, 10, 50, 100$, and 500 from up-left to right-bottom

Thus, the local magnetization reaches the equilibrium amplitude $m_s = \pm\sqrt{\gamma/g}$ when τ becomes large, whose process is described by the scaling variable (4.9) [41].

After the local magnetization is formed, the local magnetic cluster expands. Because of the term $\nabla^2 m(\boldsymbol{r})$, the wavelength of configuration changes as

$$k^2 t = \text{constant}, \tag{4.13}$$

which is called 'k^2t-scaling'. Accordingly, the typical size of cluster L increases proportionally to $\sqrt{t}$. This type of process is called 'model A' in the paper of Hohenberg and Halperin [42]. Typical pattern development is depicted in Fig. 4.1.

4.2.2 A Stretched Exponential Law for Spin-Autocorrelation Function

Below the critical temperature, the dynamics of the total magnetization is governed by this k^2t-scaling. This property causes the spin autocorrelation function in equilibrium at $H = 0$ to show a stretched exponential decay for a large t:

$$\langle \sigma_i(t)\sigma_i(0)\rangle \propto e^{-At^{\alpha}}. \tag{4.14}$$

A rough estimate of dynamics is given by considering the contribution of a large cluster [43]. Let us consider a positively ordered state below the critical temperature and a large cluster of negative spins with a radius L. The cluster shrinks from the surface, and it takes a time

$$t_L \sim L^2 \tag{4.15}$$

to disappear because of the k^2t-scaling. Thus, the average lifetime of the spins in the cluster is about t_L, which is large in a large cluster. The probability of a large cluster is given by the canonical weight with excess energy due to the surface and has a form

$$P_L \sim e^{-c\beta L^{d-1}}, \tag{4.16}$$

where c is a positive number related to the domain boundary. For large L, P_L is very small. The autocorrelation of the spin in the cluster is given by the product of the probability that the spin is in a cluster of size L, i.e., P_L and the relaxation time of the cluster t_L. The former is very small while the latter is very large. Because P_L is so small, the existence of large clusters does not contribute to static thermodynamic properties, e.g., the magnetization itself. However, for the relaxation it does. This type of non-trivial mechanism due to a product of a very small quantity and a very large quantity is called 'Griffiths-singularity' [44]. This type of singularity also appears in diluted spin systems [45].

Using the notation of the wave number $k(\sim 1/L)$,

$$P_L \to P_k \sim e^{-C\beta k^{-(d-1)}}, \tag{4.17}$$

and assuming that the relaxation of spins $C_L(t)$ in the cluster is given by

$$t_L \propto k^{-2} \to C_L(t) \sim \exp\left[-Ak^2 t\right], \tag{4.18}$$

the autocorrelation function is estimated as

$$\langle \sigma_i(t)\sigma_i(0)\rangle \sim \int d^d k P_k \exp\left[-\lambda(k)t\right] \sim \int d^d k \exp\left[-Ck^{-(d-1)} - Ak^2 t\right]. \tag{4.19}$$

Estimating this by the saddle point method, the exponent α in (4.14) was estimated as [46]

$$\alpha = \frac{d-1}{d+1}. \tag{4.20}$$

In contrast, Huse and Fisher [47] estimated it with a similar picture as

$$\alpha = \frac{d-1}{2}. \tag{4.21}$$

A Langevin type analysis supports the latter [48], while Monte Carlo simulations show the former behavior. It is still an interesting problem to clarify which case holds.

4.2.3 Conserved System: Lifshitz–Slyozov–Wigner Theory k^3t Scaling

In the case where the number of particles of each component does not change, e.g., in the segregation of two components, configuration changes only by the flow of components, i.e., by exchanges of components. This process is called 'model B' [42]. This process is expressed by the so-called Cahn–Hilliard equation. We start with the equation of continuity:

$$\frac{\partial n(\boldsymbol{r})}{\partial t} + \nabla \cdot \boldsymbol{J} = 0, \tag{4.22}$$

where $n(\boldsymbol{r})$ is the density of a component and $\boldsymbol{J}$ is the current of the component. The current is assumed to be proportional to the gradient of the chemical potential $\mu(\boldsymbol{r})$.

$$\boldsymbol{J} = -D\nabla\mu(\boldsymbol{r}). \tag{4.23}$$

The chemical potential is given by the free energy as

$$\mu(\boldsymbol{r}) = \frac{\delta F(n(\boldsymbol{r}))}{\delta n(\boldsymbol{r})}. \tag{4.24}$$

Combining them, using the same form as (4.2) replacing m by n, we have

$$\frac{\partial n(\boldsymbol{r})}{\partial t} = D\nabla^2 \left[\frac{\partial f(n(\boldsymbol{r}))}{\partial n(\boldsymbol{r})} - K(\nabla^2 n(\boldsymbol{r})) \right], \tag{4.25}$$

which is the Cahn–Hilliard equation.

In this process, segregation of components occurs. In the process, the typical size of clusters L increases proportionally to $t^{1/3}$ [49]

$$L^3 - L_0^3 \propto t - t_0, \tag{4.26}$$

where L_0 is the size at time t_0. This process is much slow compared to the non-conserved case. Here, the typical wave number k changes as

$$k^3 t = \text{constant}, \tag{4.27}$$

which is called the 'Lifshitz–Slyozov–Wagner law'. Typical pattern is depicted in Fig. 4.2. Here, the ratio of numbers of + spins and − spins is 50:50. The change

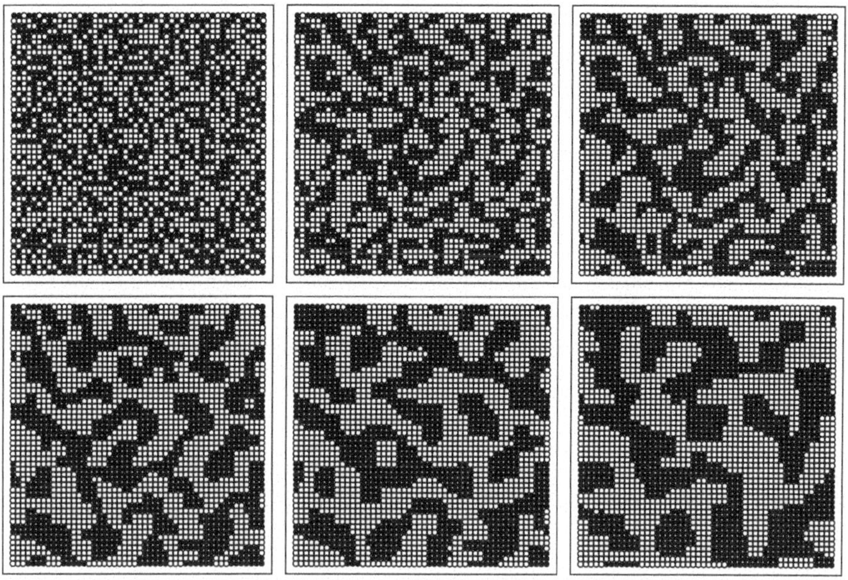

Fig. 4.2 Development of the configuration after quenching the temperature below the critical point in the conserved system. $t = 0, 10, 100, 1000, 10000$, and 100000 from up-left to right-bottom. Ratio of numbers of + spins and − spins is 50:50

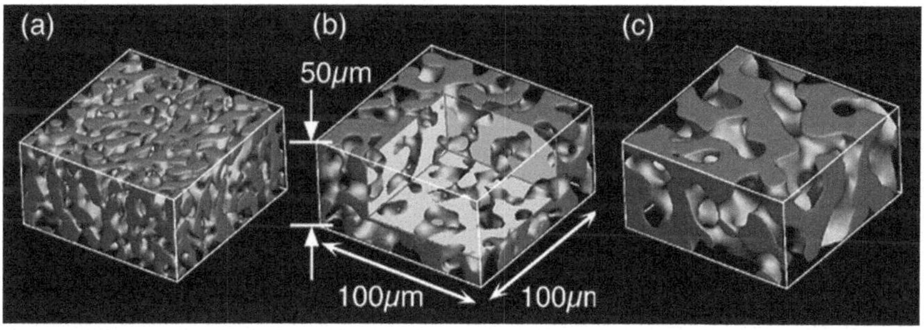

Fig. 4.3 Time evolution of interface in a binary polymer mixture at the critical composition [50]

is very slow compared to the non-conserved case (Fig. 4.1), and thus we simulated the process much longer. There, the curvature becomes smooth little by little. In principle, after a very long time, two components would segregate finally into two regions which we cannot see.

The segregation is well observed in polymer systems where the dynamics is rather slow, in which hydrodynamic flow also contributes. This type of dynamics is called 'model H' [42]. For example, the time evolution of the interface was clearly observed in a binary polymer as shown in Fig. 4.3 [50]. Solvents of block copolymer show various interesting patterns of microsegregation. For example, see the reference [51].

4.2.4 Ostwald Ripening

In the case of component ratio 30:70, the minor component is surrounded by the major one, and forms compact clusters. The evaporation rate from a small cluster is larger than that from a large one. Therefore, small clusters shrink, and large ones expand. In this way, the average size of the cluster of the minor component increases proportionally to $t^{1/3}$. This phenomenon is called Ostwald ripening [52, 53]. An example of pattern evolution is depicted in Fig. 4.4. We see the growth of the cluster sizes. Although we performed simulation 10 times longer than that in Fig. 4.2, the shapes of clusters are not smooth. If we simulate much large system for a long time, a more smooth compact cluster can be observed (not shown; please find some on the Internet).

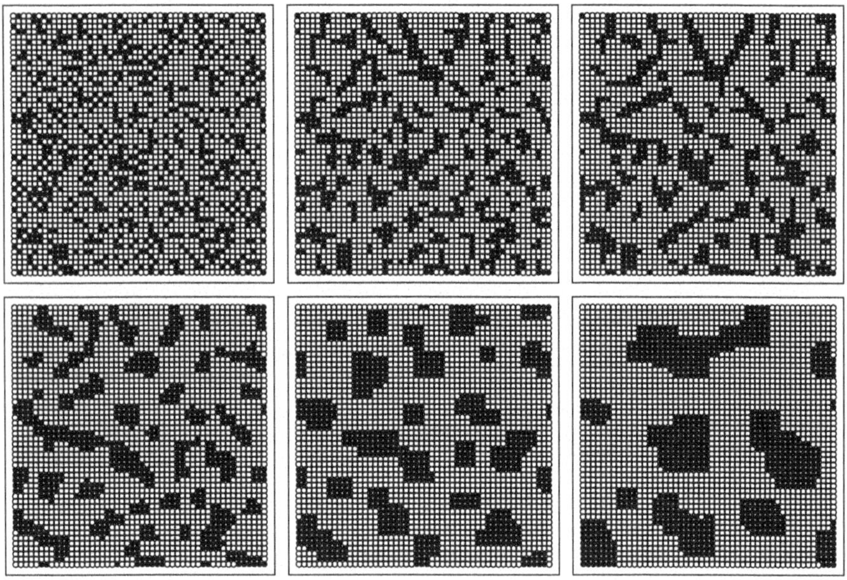

Fig. 4.4 Development of configurations after quenching the temperature below the critical point in the conserved system. $t = 0,10, 100,1000,100000$, and 1000000 from up-left to right-bottom. Ratio of number of + spins to − spins is 70:30

Part II
First-Order Phase Transition from Viewpoints of the Eigenvalue Problem

Chapter 5
Structure of Eigenvalues for the First-Order Phase Transition

5.1 Transfer Matrix

Thermodynamic properties in the equilibrium state are obtained from the partition function:

$$Z = \mathrm{Tr} e^{-\beta \mathcal{H}}. \tag{5.1}$$

The phase transition appears as a singular point of the partition function. The partition function is given by the largest eigenvalue of the transfer matrix. By analyzing the eigenvalues of the transfer matrix as a function of parameters of the Hamiltonian, e.g., the temperature and magnetic field, we find characteristics of the phase transition in the structure of eigenvalues.

We begin with a brief review of the transfer matrix method. This is a method to obtain the partition function of a one-dimensional-like system. The eigenvalues and eigenstates provide a lot of information. The largest eigenvalue is non-degenerate and gives the partition function, thus equivalently free energy. Smaller eigenvalues give information on correlation functions of various quantities. The first-order phase transition appears as a degeneracy of macroscopic phases, which corresponds to the quasi-degeneracy of the lowest two eigenvalues.

First, we review the method using a one-dimensional Ising model in zero field:

$$\mathcal{H} = -J \sum_{i=1}^{N} \sigma_i \sigma_{i+1}. \tag{5.2}$$

The partition function is given by

$$\mathrm{Tr} e^{-\beta \mathcal{H}} = \sum_{\sigma_1 = \pm 1} \sum_{\sigma_2 = \pm 1} \cdots \sum_{\sigma_N = \pm 1} e^{K\sigma_1\sigma_2 + K\sigma_2\sigma_3 + \cdots + K\sigma_{N-1}\sigma_N}, \tag{5.3}$$

S. Miyashita, *Collapse of Metastability*, Fundamental Theories of Physics 211,
https://doi.org/10.1007/978-981-19-6668-2_5

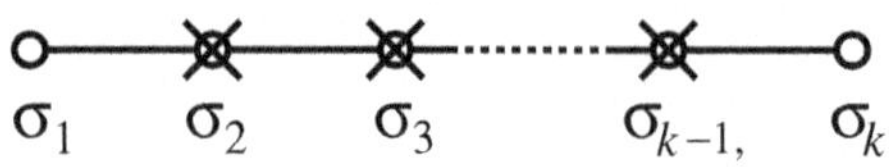

Fig. 5.1 The edge-specified partition function $Z_k(\sigma_1, \sigma_k)$ obtained by summation over all the configurations of $(\sigma_2, \ldots, \sigma_{k-1})$ keeping the edge spins σ_1 and σ_k fixed

where $K = J/k_{\mathrm{B}}T$.[1]

Here, we obtain the partition function by using the transfer matrix method. In the method, we introduce an edge-specified partition function, i.e.,

$$Z_k(\sigma_1, \sigma_k) = \sum_{\sigma_2=\pm 1} \cdots \sum_{\sigma_{k-1}=\pm 1} e^{K\sigma_1\sigma_2 + K\sigma_2\sigma_3 + \cdots + K\sigma_{k-1}\sigma_k}, \tag{5.5}$$

where the summation over all the configurations of $(\sigma_2, \ldots, \sigma_{k-1})$ is taken keeping the edge spins σ_1 and σ_k fixed. This procedure is depicted in Fig. 5.1.

Extending one more bond $K\sigma_k\sigma_{k+1}$, the edge-specified partition function $Z_{k+1}(\sigma_1, \sigma_{k+1})$ is given by

$$\begin{aligned} Z_{k+1}(+,+) &= Z_k(+,+)e^K \;\; + Z_k(+,-)e^{-K}, \\ Z_{k+1}(+,-) &= Z_k(+,+)e^{-K} + Z_k(+,-)e^K, \\ Z_{k+1}(-,+) &= Z_k(-,+)e^K \;\; + Z_k(-,-)e^{-K}, \\ Z_{k+1}(-,-) &= Z_k(-,+)e^{-K} + Z_k(-,-)e^K. \end{aligned} \tag{5.6}$$

Using the expression of the matrix product, the above relations are written as

$$\begin{pmatrix} Z_{k+1}(+,+) & Z_{k+1}(+,-) \\ Z_{k+1}(-,+) & Z_{k+1}(-,-) \end{pmatrix} = \begin{pmatrix} Z_k(+,+) & Z_k(+,-) \\ Z_k(-,+) & Z_k(-,-) \end{pmatrix} \begin{pmatrix} e^K & e^{-K} \\ e^{-K} & e^K \end{pmatrix}. \tag{5.7}$$

Using this relation iteratively, we have a compact form for the edge-specified partition function:

$$\begin{pmatrix} Z_N(+,+) & Z_N(+,-) \\ Z_N(-,+) & Z_N(-,-) \end{pmatrix} = \begin{pmatrix} e^K & e^{-K} \\ e^{-K} & e^K \end{pmatrix}^{N-1}. \tag{5.8}$$

Here, we introduce the transfer matrix

$$\hat{T} = \begin{pmatrix} e^K & e^{-K} \\ e^{-K} & e^K \end{pmatrix}. \tag{5.9}$$

[1] For this system with the open boundary condition, we can easily obtain the partition function by introducing a new variable $\tau_i = \sigma_i\sigma_{i+1}$. Each of $\{\tau_i\}$ can take ± 1 independently, and we can obtain the sum immediately as

$$Z = 2(e^K + e^{-K})^{N-1}, \tag{5.4}$$

where the factor 2 comes from the freedom of sign of $\sigma_1 (= \pm 1)$.

This matrix is diagonalized as

$$U^{-1}\begin{pmatrix} e^K & e^{-K} \\ e^{-K} & e^K \end{pmatrix} U = \begin{pmatrix} 2\cosh K & 0 \\ 0 & 2\sinh K \end{pmatrix}, \quad U = \frac{1}{\sqrt{2}}\begin{pmatrix} 1 & 1 \\ 1 & -1 \end{pmatrix}. \tag{5.10}$$

The largest eigenvalue and the second eigenvalue are

$$\lambda_1 = 2\cosh K, \quad \text{and} \quad \lambda_2 = 2\sinh K, \tag{5.11}$$

respectively. Using the relation, (5.8) is given in the form

$$\begin{pmatrix} Z_N(+,+) & Z_N(+,-) \\ Z_N(-,+) & Z_N(-,-) \end{pmatrix} = \begin{pmatrix} e^K & e^{-K} \\ e^{-K} & e^K \end{pmatrix}^{N-1} = U\begin{pmatrix} (2\cosh K)^{N-1} & 0 \\ 0 & (2\sinh K)^{N-1} \end{pmatrix} U^{-1}$$

$$= \frac{1}{2}\begin{pmatrix} (2\cosh K)^{N-1} + (2\sinh K)^{N-1}, & (2\cosh K)^{N-1} - (2\sinh K)^{N-1} \\ (2\cosh K)^{N-1} - (2\sinh K)^{N-1}, & (2\cosh K)^{N-1} + (2\sinh K)^{N-1} \end{pmatrix}. \tag{5.12}$$

The partition function of the system with an open boundary condition is given by the sum of all the spin states, i.e., $\sigma_1 = \pm 1$ and $\sigma_N = \pm 1$. Thus,

$$Z_{\mathrm{OBC}} = \sum_{\sigma_1=\pm 1}\sum_{\sigma_N=\pm 1} Z_N(\sigma_1, \sigma_N) = Z_N(+,+) + Z_N(+,-) + Z_N(-,+) + Z_N(-,-)$$

$$= 2(2\cosh K)^{N-1}, \tag{5.13}$$

which agrees with (5.4) as it should be.

For the system with a periodic boundary condition, we add one more bond $-J\sigma_N\sigma_{N+1}$ and regard $\sigma_{N+1} = \sigma_1$. In this case, the partition function is given by the sum over $\sigma_{N+1} = \sigma_1 = \pm 1$ as

$$Z_{\mathrm{PBC}} = \sum_{\sigma_1=\sigma_{N+1}=\pm 1} Z_{N+1}(\sigma_1, \sigma_{N+1}) = Z_{N+1}(+,+) + Z_{N+1}(-,-)$$

$$= (2\cosh K)^N + (2\sinh K)^N. \tag{5.14}$$

This is expressed using the matrix notation

$$Z_{\mathrm{PBC}} = \mathrm{Tr}\hat{T}^N, \tag{5.15}$$

where Tr means the trace operation for the 2×2 matrix.

As well known, the present model has no phase transition. Although any one-dimensional-like models with a shape of $W \times N$, ($W < \infty$, $N \to \infty$) have no phase transition at finite temperature, we can see a symptom of a phase transition when we increase the width of the system W, from which we study characteristics of phase transitions.

5.1.1 Ladder Systems

The transfer matrix method can be applied to one-dimensional-like models. Now we increase the width W of the system. First, we study a ladder lattice ($W = 2$) depicted in Fig. 5.2. Let the spins in the kth column be $(\sigma_{k1}, \sigma_{k2})$. Now, we obtain the partition function with the fixed edges $Z((\sigma_{11}\sigma_{12})(\sigma_{k1}\sigma_{k2}))$. As a unit of interactions of a transfer matrix, we adopt the weight of a block

$$e^{K_2\sigma_{i1}\sigma_{i2}+K_1\sigma_{i1}\sigma_{(i+1)1}+K_1\sigma_{i2}\sigma_{(i+1)2}}. \tag{5.16}$$

Then, we have the relations

$$\begin{aligned}
Z_{k+1}((++)(++)) &= Z_k((++)(++))e^{K_2}e^{K_1+K_1} + Z_k((++)(+-))e^{-K_2}e^{K_1-K_1}\\
&\quad + Z_k((++)(-+))e^{-K_2}e^{K_1-K_1} + Z_k((++)(--))e^{K_2}e^{-K_1-K_1},\\
Z_{k+1}((++)(++)) &= Z_k((++)(-+))e^{K_2}e^{K_1+K_1} + Z_k((++)(+-))e^{-K_2}e^{K_1-K_1}\\
&\quad + Z_k((++)(-+))e^{-K_2}e^{K_1-K_1} + Z_k((++)(--))e^{K_2}e^{-K_1-K_1},\\
Z_{k+1}((++)(++)) &= Z_k((++)(++))e^{K_2}e^{K_1+K_1} + Z_k((++)(+-))e^{-K_2}e^{K_1-K_1}\\
&\quad + Z_k((++)(-+))e^{-K_2}e^{K_1-K_1} + Z_k((++)(--))e^{K_2}e^{-K_1-K_1},\\
Z_{k+1}((++)(++)) &= Z_k((++)(++))e^{K_2}e^{K_1+K_1} + Z_k((++)(+-))e^{-K_2}e^{K_1-K_1}\\
&\quad + Z_k((++)(-+))e^{-K_2}e^{K_1-K_1} + Z_k((++)(--))e^{K_2}e^{-K_1-K_1},
\end{aligned} \tag{5.17}$$

which are also expressed in the form of a product of the transfer matrix:

$$\mathcal{Z}_{k+1} = \hat{T}\mathcal{Z}_k, \tag{5.18}$$

where

$$\mathcal{Z}_k \equiv \begin{pmatrix} Z_k((++)(++)) & Z_k((++)(+-)) & Z_k((++)(-+)) & Z_k((++)(--)) \\ Z_k((+-)(++)) & Z_k((+-)(+-)) & Z_k((+-)(-+)) & Z_k((+-)(--)) \\ Z_k((-+)(++)) & Z_k((-+)(+-)) & Z_k((-+)(-+)) & Z_k((-+)(--)) \\ Z_k((--)(++)) & Z_k((--)(+-)) & Z_k((--)(-+)) & Z_k((--)(--)) \end{pmatrix}, \tag{5.19}$$

and the transfer matrix $\hat{T}$ is a 4×4 matrix, which is given by

$$\hat{T} = \begin{pmatrix} e^{K_2}, 0, 0, 0 \\ 0, e^{-K_2}, 0, 0 \\ 0, 0, e^{-K_2}, 0 \\ 0, 0, 0, e^{K_2} \end{pmatrix} \begin{pmatrix} e^{2K_1} & 1 & 1 & e^{-2K_1} \\ 1 & e^{2K_1} & e^{-2K_1} & 1 \\ 1 & e^{-2K_1} & e^{2K_1} & 1 \\ e^{-2K_1} & 1 & 1 & e^{2K_1} \end{pmatrix}. \tag{5.20}$$

Here, the first matrix gives the weight of the part $e^{K_2\sigma_{i1}\sigma_{i2}}$ which is a diagonal matrix, and the second one is the weight of $e^{K_1\sigma_{i1}\sigma_{(i+1)1}+K_1\sigma_{i2}\sigma_{(i+1)2}}$. This transfer matrix $\hat{T}$ is

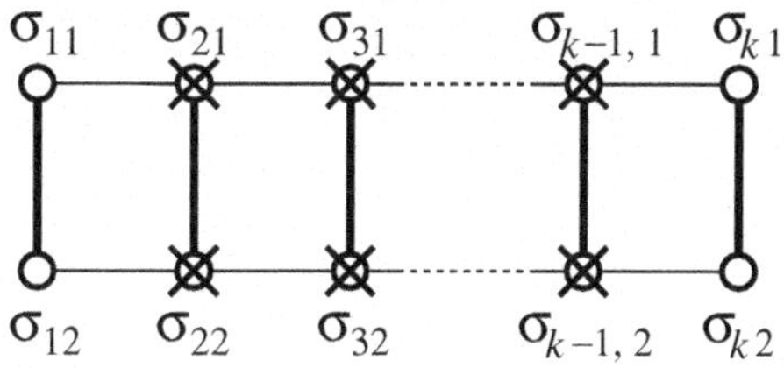

Fig. 5.2 $Z((\sigma_{11}\sigma_{12})(\sigma_{k1}\sigma_{k2}))$ for a ladder lattice. The thin lines denote bonds of K_1 and the bold line K_2

not a symmetric matrix. We may take the unit of transfer matrix in the form

$$e^{\frac{1}{2}K\sigma_{k1}\sigma_{k2}}e^{K(\sigma_{k1}\sigma_{(k+1)1}+\sigma_{k2}\sigma_{(k+1)2})}e^{\frac{1}{2}K\sigma_{(k+1)1}\sigma_{(k+1)2}}. \tag{5.21}$$

In this case, the transfer matrix is given in a symmetric form:

$$\mathcal{T}=\begin{pmatrix} e^{\frac{1}{2}K_2},0,0,0 \\ 0,e^{-\frac{1}{2}K_2},0,0 \\ 0,0,e^{-\frac{1}{2}K_2},0 \\ 0,0,0,e^{\frac{1}{2}K_2}\end{pmatrix}\begin{pmatrix} e^{2K_1} & 1 & 1 & e^{-2K_1} \\ 1 & e^{2K_1} & e^{-2K_1} & 1 \\ 1 & e^{-2K_1} & e^{2K_1} & 1 \\ e^{-2K_1} & 1 & 1 & e^{2K_1}\end{pmatrix}\begin{pmatrix} e^{\frac{1}{2}K_2},0,0,0 \\ 0,e^{-\frac{1}{2}K_2},0,0 \\ 0,0,e^{-\frac{1}{2}K_2},0 \\ 0,0,0,e^{\frac{1}{2}K_2}\end{pmatrix} \tag{5.22}$$

which is convenient for analytical and also numerical treatments.

When the width W becomes large, a row contains W spins:

$$(\sigma_{i1},\sigma_{i2})\rightarrow(\sigma_{i1},\sigma_{i2},\ldots\sigma_{iW}), \tag{5.23}$$

and the dimension of the transfer matrix becomes 2^W. When W increases, the lattice becomes like a two-dimensional lattice, and shows a symptom of phase transition. In the limit $W\rightarrow\infty$, the lattice becomes a two-dimensional lattice, and we obtain the partition function of the two-dimensional Ising model [54].

5.1.2 *Free Energy*

The free energies per a spin ($f=-(1/N)k_BT\ln Z$) for the systems of $W=1$ with the open boundary condition and the periodic boundary conditions are given by

$$f_{\mathrm{OBC}}=-\frac{1}{N}k_BT\ln\left(2(2\cosh K)^{N-1}\right)=-k_BT\left(\frac{N-1}{N}\ln(2\cosh K)+\frac{1}{N}\ln 2\right) \tag{5.24}$$

and

$$f_{\mathrm{PBC}}=-\frac{1}{N}k_BT\ln\left((2\cosh K)^N+(2\sinh K)^N\right)$$

$$= -k_{\mathrm{B}}T\left(\ln(2\cosh K) + \frac{\ln(1+(\tanh K)^N)}{N}\right), \tag{5.25}$$

respectively. Although the expressions are different, in the thermodynamic limit ($N \to \infty$), the free energy is given by

$$f = -\lim_{N\to\infty}\frac{1}{N}k_{\mathrm{B}}T\ln Z(T) = -k_{\mathrm{B}}T\ln(2\cosh K) = -k_{\mathrm{B}}T\ln\lambda_1. \tag{5.26}$$

Namely, the free energy is given by the largest eigenvalue of the transfer matrix.

Above, we considered the model with only two eigenstates, but this relation generally holds. For a system of length N and width W with the periodic boundary condition, the partition function is given by

$$Z_{\mathrm{PBC}} = \mathrm{Tr}\hat{T}^N = \sum_{i=1}^{2^W}\lambda_i^N, \tag{5.27}$$

where $\{\lambda_i\}$ $(i = 1, \ldots 2^W)$ are the transfer eigenvalues of $\hat{T}$:

$$\hat{T}|i\rangle = \lambda_i, |i\rangle \quad \lambda_1 > \lambda_2 \geq \lambda_3 \cdots. \tag{5.28}$$

The transfer matrix is a positive matrix, that is, all the matrix elements are positive. In general, the positive matrix has one maximum eigenvalue (non-degenerate), and absolute values of all the eigenvalues are less than the maximum one (Perron–Frobenius theory).

Then, the free energy per spin in the limit $N \to \infty$ is

$$f(T) = -\lim_{N\to\infty}\frac{1}{N}k_{\mathrm{B}}T\ln\left(\lambda_1^N + \lambda_2^N + \cdot + \lambda_{2^W}^N\right)$$

$$f(T) = -k_{\mathrm{B}}T\ln\lambda_1 - \frac{1}{N}k_{\mathrm{B}}T\ln\left(1 + \left(\frac{\lambda_2}{\lambda_1}\right)^N + \left(\frac{\lambda_3}{\lambda_1}\right)^N + \cdots\right) \simeq -k_{\mathrm{B}}T\ln\lambda_1. \tag{5.29}$$

Thus, the free energy per spin is given by the largest eigenvalue of the transfer matrix λ_1. Because of non-degeneracy of the maximum eigenvalue,

$$\lim_{N\to\infty}\left(\frac{\lambda_2}{\lambda_1}\right)^N = 0, \tag{5.30}$$

which ensures the uniqueness of the equilibrium state. It should be noted that when the system is a coexistence state where different ordered states coexist, the eigenvalues degenerate in the limit $N \to \infty$, although the maximum eigenstate is always non-degenerate for $N < \infty$.

5.1.3 Correlation Functions

The correlation function of spins is given by the higher eigenvalues. The correlation function of a quantity O at positions in the ith unit and the jth unit is given by

$$\langle O(i)O(j)\rangle = \frac{\sum_{\text{all the states}} e^{-\beta\mathcal{H}_1}\cdots e^{-\beta\mathcal{H}_i} O e^{-\beta\mathcal{H}_{i+1}}\cdots e^{-\beta\mathcal{H}_j} O e^{-\beta\mathcal{H}_{j+1}}\cdots e^{-\beta\mathcal{H}_L}}{\sum_{\text{all the states}} e^{-\beta\mathcal{H}_1} e^{-\beta\mathcal{H}_2}\cdots e^{-\beta\mathcal{H}_L}}$$

$$= \frac{\mathrm{Tr}\hat{T}^i O \hat{T}^{j-i} O \hat{T}^{L-j}}{\mathrm{Tr}\hat{T}^L}, \tag{5.31}$$

where $\mathcal{H}_i$ is a unit of Hamiltonian for the transfer matrix $\hat{T}$. Here, we assume that $\langle 1|O|1\rangle = 0$. If $\langle 1|O|1\rangle$ is not zero, we redefine $O \to O - \langle 1|O|1\rangle$.

Then, assuming $L \gg |j-i|$, the correlation function is given by

$$\langle O(i)O(j)\rangle \simeq \frac{\sum_{k=2}^{2^M} |\langle 1|O|k\rangle|^2 \lambda_k^{|i-j|} \times \lambda_1^{L-(|i-j|)}}{\lambda_1^L} \simeq \sum_{k=2}^{2^M} |\langle 1|O|k\rangle|^2 \left(\frac{\lambda_k}{\lambda_1}\right)^{|i-j|}, \tag{5.32}$$

because $\lambda_1 > \lambda_i$ ($i \geq 2$). Here, $\langle 1|O|k\rangle$ is the matrix element of the quantity O for the eigenstate of the largest eigenvalue $|1\rangle$ and the kth eigenstate.

Thus, the correlation function at a long distance is mainly given by λ_{k_0}/λ_1 where λ_{k_0} is the largest eigenstate with nonzero matrix element ($\langle 1|O|k_0\rangle \neq 0$).

$$\langle O(i)O(j)\rangle \simeq Ae^{-|i-j|/\xi}, \quad A = |\langle 1|O|k_0\rangle|^2, \quad \xi_k^{-1} = \left|\ln\left(\frac{\lambda_{k_0}}{\lambda_1}\right)\right|. \tag{5.33}$$

At the critical point, the correlation length diverges, which indicates

$$\frac{\lambda_2}{\lambda_1} \to 1. \tag{5.34}$$

That is, when the system approaches the critical point, λ_2 comes close to λ_1. At the critical temperature, the difference of λ_1 and λ_2 becomes small with a power of system size N. Below the critical temperature, reflecting the appearance of a long-range order, the spin correlation length diverges. There, λ_1 and λ_2 are quasi-degenerate, i.e., the difference becomes small exponentially with the system size N, and they degenerate in the thermodynamic limit.

5.1.4 Temperature Dependence of the Eigenvalues

As an example, the temperature dependence of eigenvalues $\{\lambda_i\}$ of the transfer matrix of $W = 10$ at $H = 0$ are depicted in Fig. 5.3, where $\{-\ln \lambda_i\}$ (corresponding to the free energy) instead of $\{\lambda_i\}$ are depicted as a function the temperature T.

At high temperatures, λ_1 and λ_2 are well separated. As the temperature decreases, the separation decreases and almost disappears around the critical temperature $T_C \simeq 2.269$ ($K_c \simeq 0.4407$). It is known that the correlation function of spin correlation function ξ_1 decreases as

$$\xi_1 \propto |T - T_C|^{-\nu}, \tag{5.35}$$

where ν is the critical exponent of the correlation length. For the two-dimensional Ising model, $\nu = 1$. By a scaling relation, we know that at $T = T_C$, the difference of λ_1 and λ_2 decreases with the size as

$$\lambda_1 - \lambda_2 \propto W^{-1/\nu}. \tag{5.36}$$

Below the critical temperature ($T < T_C$), the separation decreases exponentially with W

$$\lambda_1 - \lambda_2 \propto e^{-cW} \quad c > 0. \tag{5.37}$$

(W dependence is not shown.) This quasi-degeneracy of λ_1 and λ_2 indicates the spontaneous breakdown of the symmetry, i.e., appearance of the spontaneous magnetization. This degeneracy of λ_1 and λ_2 is a characteristic of the coexistence of different states in the thermodynamic limit from the viewpoint of an eigenvalue problem of

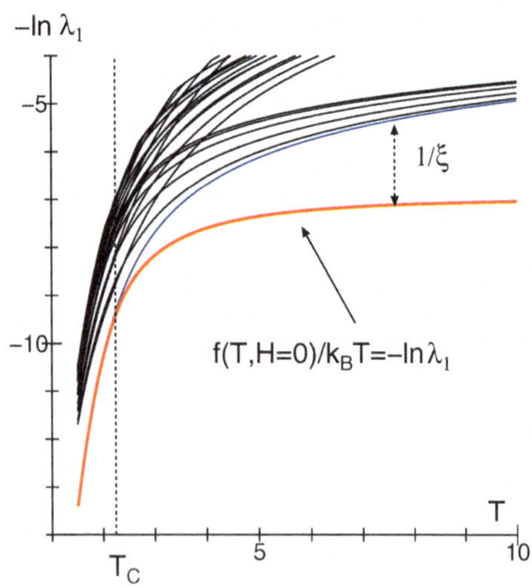

Fig. 5.3 Temperature dependence of eigenvalues ($\{-\ln \lambda_i\}$) as a function of the temperature T of a system of $W = 10$ with $H = 0$

the transfer matrix. That is, the first-order phase transition is characterized by this degeneracy.

5.1.5 *Field Dependence of the Eigenvalues Below the Critical Temperature*

Next, we study the field dependence of the eigenvalues on the magnetic field in the ordered state. The eigenvalues at $T = 2.0 < T_{\mathrm{C}}$ are depicted as a function of the field in Fig. 5.4. The lowest curve gives the ground state ($|1\rangle$) which is an almost perfectly ordered state. The magnetization of the state per site m is given by the slope of the curve, i.e.,

$$m = -\frac{\partial f}{\partial H} = -k_{\mathrm{B}}T\frac{\partial(-\ln\lambda_1)}{\partial H}. \tag{5.38}$$

The lowest branches show a sharp cross at $H = 0$. The branch which is regarded as a continuation of λ_1 after the cross gives a state with the magnetization in the opposite direction to the field, and it represents a metastable state. There is a very small gap given by (5.37) at the crossing point ($H = 0$). Namely, λ_1 and λ_2 form an avoided level crossing, which is also called quasi-crossing. This avoided level crossing is a characteristic of the first-order phase transition. When we sweep the field slowly, the equilibrium state of the system given by $|1\rangle$ moves following the lowest branch, which is the isothermal change of state. But at the crossing, the state tends to stay in the continuation, i.e., the state $|2\rangle$, which has the same magnetization, unless the sweeping velocity is extremely small. Thus, in a realistic velocity, after the crossing at $H = 0$, the system stays in the almost same state as the metastable state.

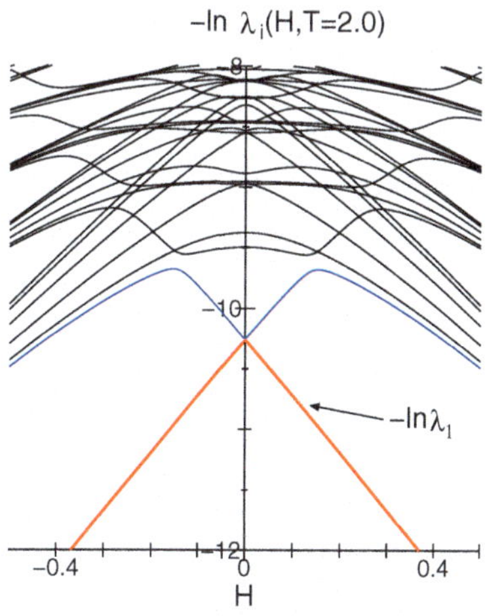

Fig. 5.4 Field dependence of eigenvalues of a system of $W = 10$ at $T = 2.0 < T_{\mathrm{C}} \simeq 2.269$

5.2 Eigenvalue Analysis of Dynamical Processes

Properties of the relaxation process of the metastable state are studied as an eigenvalue problem of the time-evolution operator. The time-evolution operator may be called 'Liouvillian' and is denoted by $\mathcal{L}$ or L.[2] The Liouvillian for a stochastic process is generally given by the master equation (see Appendix 8.2). Let us begin with a brief review of the dynamics of the probability distribution of states by the master equation.

5.2.1 *Eigenstates of Master Equation*

The time evolution of probability of the ith state $P(i,t)$ is given by the transition probabilities $(\{\omega_{i\to j}\})$ as explained in Appendix 8.2 (8.54):

$$P(i,t+\Delta t) = P(i,t) - \sum_{j\neq i} P(i,t)\omega_{i\to j}\Delta t + \sum_{j\neq i} P(j,t)\omega_{j\to i}\Delta t, \tag{5.39}$$

and in the vector expression of the probability distribution $\boldsymbol{P}(t)$

$$\boldsymbol{P}(t+\Delta t) = \mathcal{L}\boldsymbol{P}(t). \tag{5.40}$$

To study the time evolution of $\boldsymbol{P}(t)$, let the eigenvalues and the corresponding eigenstates of $\mathcal{L}$ be $\hat{\lambda}_m$ and $|\phi_m\rangle$:

$$\mathcal{L}|\phi_m\rangle = \hat{\lambda}_m|\phi_m\rangle, \quad m = 1, 2, \ldots, \mathcal{M}. \tag{5.41}$$

This matrix is not symmetric, and the right-eigenstate and left-eigenstate are different. As to the symmetrization of the time-evolution operator, we will explain the procedure in Sect. 8.2.3. Generally, $\hat{\lambda}_m$ is complex, but if L can be symmetrized, the eigenvalues are real. In this section, we consider the cases with real eigenvalues.

The time evolution of the probability is given by the eigenstates $|\phi_m\rangle$, $(m = 1, \ldots, \mathcal{M})$. The initial probability $\boldsymbol{P}(0)$ is decomposed as

$$\boldsymbol{P}(0) = \sum_{m=1}^{\mathcal{M}} c_m|\phi_m\rangle. \tag{5.42}$$

Thus, the probability at a time $t = n\Delta t$ is expressed as

$$\boldsymbol{P}(t) = \mathcal{L}^n\boldsymbol{P}(0) = \sum_{m=1}^{\mathcal{M}} c_m\hat{\lambda}_m^n|\phi_m\rangle. \tag{5.43}$$

---

[2] It is the same symbol as the Lagrangian. Do not confuse.

5.2.2 Approach to the Stationary State

If the condition that all states can be transferred to all the states in the time evolution, the process is called 'ergodic'. In this case, at some n, all the matrix elements are positive:

$$\left(\mathcal{L}^n\right)_{ij} > 0. \tag{5.44}$$

Thus, in this case, the time-evolution matrix is the positive matrix, and the Perron–Frobenius theorem tells that the maximum eigenvalue $\hat{\lambda}_1$ is positive and non-degenerate (the so-called Perron root), and all the elements of its eigenstate $|\phi_1\rangle$ are positive. And the absolute values of other eigenvalues are less than λ_1:

$$\hat{\lambda}_1 > |\hat{\lambda}_k|, \quad k \geq 2. \tag{5.45}$$

Thus, $\boldsymbol{P}(t)$ given in (5.43) converges to $|\phi_1\rangle$:

$$\boldsymbol{P}(N\Delta t) = \sum_{m=1}^{\mathcal{M}} c_m \hat{\lambda}^N |\phi_m\rangle \to c_1 \hat{\lambda}_1^N |\phi_1\rangle. \tag{5.46}$$

Thus, $|\phi_1\rangle$ is the stationary state of the master equation. The normalization must be kept,

$$\sum_{m=1}^{\mathcal{M}} P(m,t) = 1 \tag{5.47}$$

at any time t. Therefore, the following properties must be satisfied.

$$c_1 = 1, \quad \hat{\lambda}_1 = 1 \tag{5.48}$$

and

$$\sum_{k=1}^{\mathcal{M}} \phi_{1k} = 1, \tag{5.49}$$

where ϕ_{1k} is the kth component of the eigenvector $|\phi_1\rangle$. From this observation, we find that all the probability distributions $\boldsymbol{P}(t)$ must contain $|\phi_1\rangle$ with the coefficient $c_1 = 1$. Thus, the stationary state is given by

$$\boldsymbol{P}_{\mathrm{st}} = |\phi_1\rangle, \tag{5.50}$$

which satisfies

$$\mathcal{L}\boldsymbol{P}_{\text{st}} = \boldsymbol{P}_{\text{st}}. \tag{5.51}$$

The contribution to (5.47) from other eigenstates $c_m\hat{\lambda}_m^n|\phi_m\rangle$ must be zero independently of n. Thus, the following property must hold:

$$\sum_k \phi_{mk} = 0, \quad m > 1. \tag{5.52}$$

This property indicates that the contribution of $|\phi_m\rangle$ causes an increase of some part of $\boldsymbol{P}_{\text{st}}$ and decrease of the other part. Thus, we regard $|\phi_m\rangle(m > 1)$ as a mode of deformation of the probability distribution of the stational state. The amplitude of the deformation decreases as λ_k^n, which gives the relaxation of the mode of deformation (5.54). Thus, there are $\mathcal{M} - 1$ relaxation modes in the system.

We define relaxation time for the mth eigenstate. Because $|\lambda_k| < \lambda_1 = 1$, $(k \geq 2)$, we can write (5.43) as

$$\boldsymbol{P}(t) = \mathcal{L}^n\boldsymbol{P}(0) = \sum_{m=1}^{\mathcal{M}} c_m\hat{\lambda}_m^n|\phi_m\rangle = \sum_{m=1}^{\mathcal{M}} c_m e^{-|\ln\hat{\lambda}_m|t/\Delta t}|\phi_m\rangle. \tag{5.53}$$

We define relaxation time for the mth eigenstate as

$$\hat{\tau}_m = \frac{\Delta t}{|\ln\hat{\lambda}_m|}. \tag{5.54}$$

Then, $\boldsymbol{P}(t)$ is given in the form:

$$\boldsymbol{P}(t) = \sum_{m=1}^{\mathcal{M}} c_m e^{-t/\hat{\tau}_m}|\phi_m\rangle = \boldsymbol{P}_{\text{st}} + \sum_{m=2}^{\mathcal{M}} c_m e^{-t/\hat{\tau}_m}|\phi_m\rangle. \tag{5.55}$$

In the differential form explained in Appendix 8.2 (8.66),

$$\frac{d}{dt}\boldsymbol{P}(t) = L\boldsymbol{P}(t), \quad L = \frac{\mathcal{L}-1}{\Delta t}, \quad L|\phi_m\rangle = \lambda_m|\phi_m\rangle, \quad \lambda_k = \frac{\hat{\lambda}_k - 1}{\Delta t},$$

the time evolution of $\boldsymbol{P}(t)$ is given by

$$\boldsymbol{P}(t) = \sum_{m=1}^{\mathcal{M}} c_m e^{-\lambda_m t}|\phi_m\rangle \equiv \sum_{m=1}^{\mathcal{M}} c_m e^{-t/\tau_m}|\phi_m\rangle. \tag{5.56}$$

Thus, $\hat{\tau}_k = \tau_k$ and hereafter, we write $\hat{\tau}_k$ simply by τ_k. Comparing to (5.55),

$$\lambda_m = \frac{1}{\hat{\tau}_m} = \frac{|\ln \hat{\lambda}_k|}{\Delta t}. \tag{5.57}$$

5.3 Kinetic Ising Model

For dynamics of the Ising spin system

$$\mathcal{H}(\sigma_1, \sigma_2, \ldots, \sigma_N) = -J \sum_{<i,j>} \sigma_i \sigma_j - H \sum_{i=1}^{N} \sigma_i, \tag{5.58}$$

we adopt the kinetic Ising model which was explained in Sect. 3.4.1.

In the non-conserved systems in which each spin can flip independently and the total magnetization is not conserved, we adopt the single spin flip update, where the transition probability is given for a flip of the spin σ_k which changes the state from the state i to the state j:

$$i = (\sigma_1, \ldots, \sigma_k, \ldots, \sigma_N), \quad j = (\sigma_1, \ldots, -\sigma_k, \ldots, \sigma_N). \tag{5.59}$$

For the transition probability, we adopt the Glauber model (3.18) in which the transition probability of a single spin flip of the spin σ_k per unit time is given by

$$w_{(\sigma_1,\ldots,\sigma_k,\ldots,\sigma_N)\to(\sigma_1,\ldots,-\sigma_k,\ldots,\sigma_N)} = \frac{e^{-\beta\mathcal{H}(\sigma_1,\ldots,-\sigma_k,\ldots,\sigma_N)}}{e^{-\beta\mathcal{H}(\sigma_1,\ldots,\sigma_k,\ldots,\sigma_N)} + e^{-\beta\mathcal{H}(\sigma_1,\ldots,-\sigma_k,\ldots,\sigma_N)}}, \tag{5.60}$$

which satisfies the detailed balance (3.16)(= (8.71)): $w_{i\to j} P_{\mathrm{eq}}(i) = w_{j\to i} P_{\mathrm{eq}}(j)$. When the transition probability satisfies the detailed balance, the canonical distribution $\boldsymbol{P}_{\mathrm{eq}}$ is the stationary state: i.e.,

$$\mathcal{L}\boldsymbol{P}_{\mathrm{eq}} = \boldsymbol{P}_{\mathrm{eq}}. \tag{5.61}$$

In the conserved systems in which the total magnetization is conserved, we adopt the update of the state by exchange of spins in a spin pair with opposite signs $\sigma_k = -\sigma_l$, which changes the state from the state i to the state j:

$$i = (\sigma_1, \ldots, \sigma_k, \ldots, \sigma_l, \ldots, \sigma_N), \quad j = (\sigma_1, \ldots, \sigma_l, \ldots, \sigma_k, \ldots, \sigma_N). \tag{5.62}$$

The transition probability is given for the exchange of the spins (σ_k, σ_l) which is called 'Kawasaki dynamics'.

5.3.1 Demonstration in a Small System of 2 × 2 System

First, we demonstrate the eigenvalue problem of a small system of 2 × 2 (Fig. 5.5), and see how the eigenvalues and eigenstates behave explicitly. Here, we adopt the periodic boundary conditions in both directions. The time-evolution operator is given by a 16 × 16 matrix, and we use the bases states representing the spin configuration $(\sigma_1, \sigma_2, \sigma_3, \sigma_4)$ as $(-, -, -, -), (-, -, -, +), \ldots, (+, +, +, +)$. The spins, $\sigma_1, \sigma_2, \sigma_3$ and σ_4, locate at (1,1),(1,2),(2,1), and (2,2), respectively. A number '*state*' is assigned to denote the spin configuration $(\sigma_1, \sigma_2, \sigma_3, \sigma_4)$ by

$$state = 1 + b_1 + 2b_2 + 4b_3 + 8b_4, \tag{5.63}$$

where $b_i = (\sigma_i + 1)/2$.

5.3.1.1 Temperature Dependence of Eigenvalues

Temperature dependences of $1/\tau_i (= \lambda_i)$ are depicted in Fig. 5.6a. We find the second lowest eigenvalue comes close to zero (λ_1) at low temperatures which indicates a very long relaxation time between states $(+, +, +, +)$ and $(-, -, -, -)$.

In Fig. 5.6b, components of eigenstates $|P_k(state)|$ for a few eigenstates k =1,2,3,4, and 16 at $T/J = 3$ are plotted. While the eigenstate of $k = 1$ is the stationary state and

$$\sum_{state=1}^{16} P_k(state) = 1. \quad k = 1, \tag{5.64}$$

the eigenstates of $k > 1$ give modes of modification of the probability distribution and

$$\sum_{state=1}^{16} P_k(state) = 0. \quad k > 1. \tag{5.65}$$

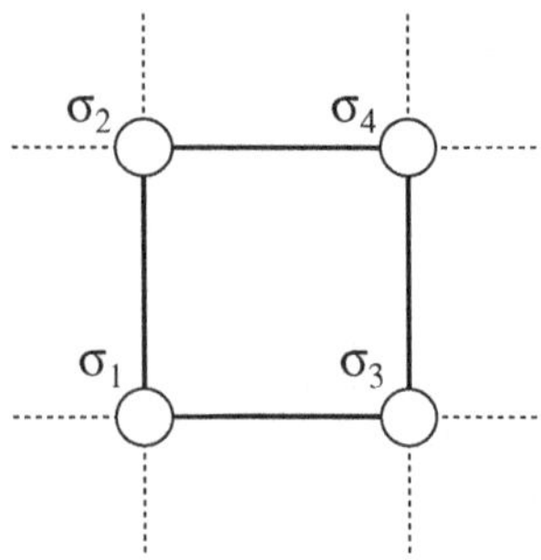

Fig. 5.5 A lattice of 2 × 2 system with the periodic boundary conditions

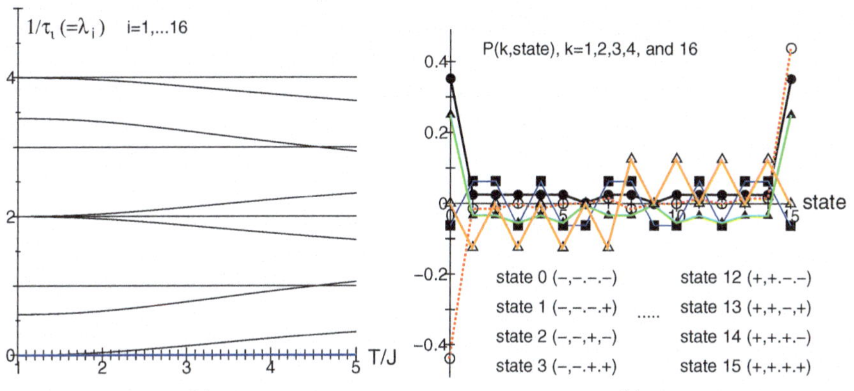

Fig. 5.6 **a** Temperature dependence of $1/\tau_i(=\lambda_i)$ for the kinetic Ising model of the Glauber dynamics for the 2×2 system (Fig. 5.5) at $H = 0$. **b** Eigenstates $\boldsymbol{P}_k$ for k =1,2,3,4, and 16 at $T/J = 3$. The states are normalized to be $\sum_{k=1}^{16} |P_i(k)| = 1$. The state denotes the spin configuration $(\sigma_1, \sigma_2, \sigma_3, \sigma_4)$ by the number $state = 1 + b_1 + 2b_2 + 4b_3 + 8b_4$, where $b_i = (\sigma_i + 1)/2$

Therefore, in the figure, we plot amplitudes of $P_k(state)$ for $k > 1$ normalized as

$$\sum_{state=1}^{16} |P_k(state)| = 1. \tag{5.66}$$

The probability distributions $P_1(state)$, $(state = 1, \ldots 8)$ of the equilibrium state $(\lambda_1 = 0)$ are plotted by closed circles connected by a bold line. We find that the weights of $(+, +, +, +)$ and $(-, -, -, -)$ are large reflecting the ferromagnetic ordering. The second eigenstate, i.e., $P_2(state)$ for λ_2, is plotted by open circles connected by a dotted line which gives a relaxation mode to change between $(+, +, +, +)$ and $(-, -, -, -)$. The mode of the fastest relaxation time (λ_{16}) is plotted by open triangles.

In Fig. 5.6a, we find flat lines of integer values. These modes indicate that the configuration has sites which can change freely, i.e., the change of energy by the flip is zero $\Delta E = 0$.

$$w_{(\sigma_1,\ldots,\sigma_i,\ldots,\sigma_N)\to(\sigma_1,\ldots,-\sigma_i,\ldots,\sigma_N)} = 1/2. \tag{5.67}$$

The flat lines of $\tau = 2$, 3, and 4 denote the states which have several free flip spin sites. This type of fast relaxation is not important for metastability. The property of the metastable state is given by small eigenstates. In Fig. 5.6a, only one lowest mode $(k = 2)$ relates to the metastability.

5.3.1.2 Field Dependence of Eigenvalues

Next, we study field dependence. Field dependences of $1/\tau_i$ for a 2×2 system with the periodic boundary conditions in both directions are depicted in Fig. 5.7a at $T/J = 2$. In Fig. 5.7b, the eigenstates $\boldsymbol{P}_k$ for k =1,2,3,4, and 16 at $H/J = 2$ are plotted. The amplitudes are normalized to be $\sum_{state=1}^{16} |P_k(state)| = 1$ in the same manner of Fig. 5.6b. We find the equilibrium state ($\lambda_1 = 0$) has a large weight at $(+, +, +, +)$ because of the positive field $H/J = 2$. But, reflecting the metastability of the state $(-, -, -, -)$, the second eigenstate (λ_2), i.e., the slowest relaxation mode, gives a mode to change between $(+, +, +, +)$ and $(-, -, -, -)$, which represents the relaxation mode of metastability of the state $(-, -, -, -)$.

5.3.2 *Master Equation for the Magnetization for a Model with Long-Range Interaction*

In principle, to study the characteristics of phase transition, we have to study the size dependence of quantities to clarify the singularity due to the phase transition. However, direct study of the eigenvalue structure of systems of large lattices is difficult for the kinetic Ising model with short-range interaction because the dimension of the time-evolution operator increases very rapidly, i.e., increases as 2^N. An attempt to study a larger lattice with short-range interaction is given in the next Sect. 5.3.3.

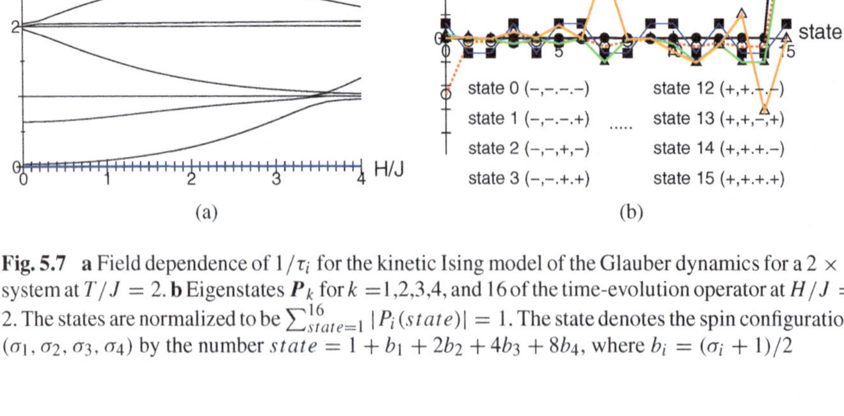

Fig. 5.7 **a** Field dependence of $1/\tau_i$ for the kinetic Ising model of the Glauber dynamics for a 2×2 system at $T/J = 2$. **b** Eigenstates $\boldsymbol{P}_k$ for k =1,2,3,4, and 16 of the time-evolution operator at $H/J = 2$. The states are normalized to be $\sum_{state=1}^{16} |P_i(state)| = 1$. The state denotes the spin configuration $(\sigma_1, \sigma_2, \sigma_3, \sigma_4)$ by the number $state = 1 + b_1 + 2b_2 + 4b_3 + 8b_4$, where $b_i = (\sigma_i + 1)/2$

In this circumstance, we study the eigenvalue structure of large systems in a mean-field type model. Here, we adopt the long-range interacting system, i.e., the Husimi–Temperley model (8.32), in which the states are characterized by the magnetization M:

$$\mathcal{H}_{\rm HT} = \frac{Jz}{N}\sum_i\sum_j \sigma_i\sigma_j - H\sum_{i=1}^N \sigma_i = \frac{Jz}{N}M^2 - HM, \quad M = \sum_{i=1}^N \sigma_i. \tag{5.68}$$

This model corresponds to the mean-field model as we see Sect. 8.1.4, and we do not need to consider explicitly the spin configuration $(\sigma_1, \sigma_2, \ldots, \sigma_N)$.

In the model, we do not have inhomogeneity of spin structure in space, and the model does not show nucleation phenomena. But it still shows important characteristics of the dynamics of the phase transition. It should be noted that the model (5.68) is not only an artificial tactful model, but also it governs the nature of some realistic phase transitions in systems with long-range interaction, e.g., the spin-crossover systems (Sect. 2.4.3) with elastic interaction [20, 21].

As mentioned in Sect. 3.4.1, the states of the present model are characterized by the magnetization M ($M = -N, -N+2, \ldots, N$.). If we express the states of the system by M, we consider the probability $P(M)$ that the system has the magnetization M. Then, the number of states is $N+1$ in contrast to 2^N in the original model. By a single spin flip ($\sigma_k \to -\sigma_k$), M changes by 2, and taking into account the degeneracy $D(M)$, the master equation for the probability of state with magnetization M at a time t, $P(M,t)$, is given by (3.23) as explained in Sect. 3.4.1. The transition probability of the process $M \to M-2$ of the Glauber type is given by (3.24), and the master equation is given explicitly by (3.25):

$$\begin{aligned}
\tau_0\frac{\partial P(M,t)}{\partial t} &= -\frac{N+M}{2}\frac{e^{\beta(Jz(M-1)/N+H)}}{2\cosh(\beta(Jz(M-1)/N+H))}P(M,t) \\
&\quad -\frac{N-M}{2}\frac{e^{\beta(Jz(M+1)/N+H)}}{2\cosh(\beta(Jz(M+1)/N+H))}P(M,t) \\
&\quad +\frac{N+M+2}{2}\frac{e^{\beta(Jz(M+1)/N+H)}}{2\cosh(\beta(Jz(M+1)/N+H))}P(M+2,t) \\
&\quad +\frac{N-M+2}{2}\frac{e^{\beta(Jz(M-1)/N+H)}}{2\cosh(\beta(Jz(M-1)/N+H))}P(M-2,t),
\end{aligned}$$

which was studied by a Fokker–Planck equation in Sect. 3.4.1.

For the system with N spins, this process is expressed by a matrix of the time-evolution operator (a $(N+1)\times(N+1)$ matrix). We obtain the eigenvalues and eigenstates of the matrix. In the present case, we can manage systems up to large values of N.

5.3.2.1 Temperature Dependence of Eigenvalues

In Fig. 5.8, the temperature dependence of eigenvalues $\{\lambda_i\}$ of the time-evolution operator with $N = 100, 200$ and 400 are depicted. Above $T_C(= J)$, the relaxation time is given by

$$\tau = \frac{\tau_0}{\lambda_2 - \lambda_1}, \quad \lambda_1 = 0. \tag{5.69}$$

As the temperature comes close to T_C, λ_2 becomes small, which indicates that the relaxation time diverges at the critical point, that is, the critical slowing down.

Below T_C, the second eigenvalue λ_2 is almost degenerate to $\lambda_1(= 0)$ indicating the appearance of the spontaneous magnetization. For finite size systems, τ_0/λ_2 gives the times scale of a plateau in the motion of magnetization $m(t)$. The relaxation time of fluctuations of the magnetization in the symmetry broken state is given by $1/\lambda_3$. λ_3 remains finite in the thermodynamic limit while λ_2 becomes zero in the limit. λ_3 $(T < T_C)$ approaches 0 at T_C as N increases, which reflects the fact that the relaxation time τ_0/λ_3 diverges at the critical point.

5.3.2.2 Field Dependence of Eigenvalues

Next, we study field dependence of the eigenvalues. Below T_C, there exists a spinodal point H_{SP}. The spinodal point is given by

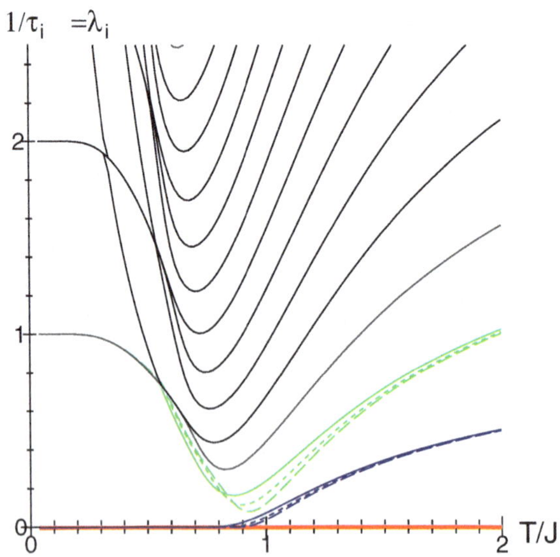

Fig. 5.8 Temperature dependence of eigenvalues of the kinetic Ising model (Husimi–Temperley model with $z = 1$) for $N = 100$ are plotted by solid lines. The critical temperature of this model is $T_C/J = 1$. The short dashed lines show λ_2 and λ_3 for $N = 200$. The dash lines show λ_2 and λ_3 for $N = 400$

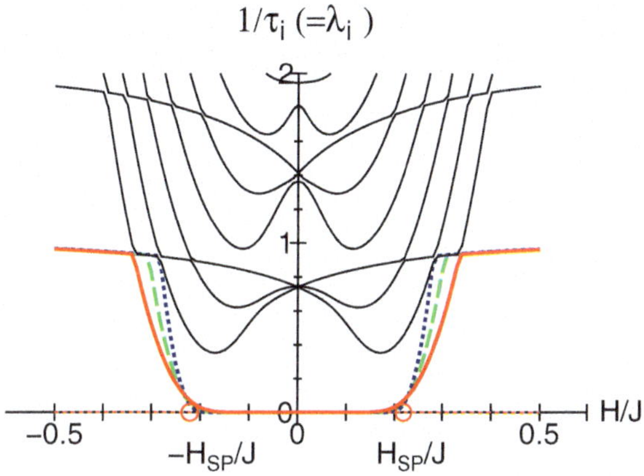

Fig. 5.9 Field dependence of eigenvalues of the kinetic Ising model (Husimi–Temperley model with $z = 1$). $K = (1/k_B T) = 1.8$ for $N = 100$. For, $N = \infty$, the spinodal point is $H_{sp} \sim 0.22$ at this temperature. The second smallest eigenvalues for $N = 200$, and 400 are plotted by dashed and doted lines, respectively. The circles denote the spinodal points in the limit $N \to \infty$

$$H_{SP} = \pm JzX \mp \frac{k_B T}{2} \ln \left(\frac{1+X}{1-X} \right), \quad X = \sqrt{1 - \frac{k_B T}{Jz}}. \tag{5.70}$$

Below the spinodal point ($|H| < |H_{SP}|$), the metastable state survives where λ_2 is nearly 0.

In Fig. 5.9, the field dependence of eigenvalues at $T/J = 5/9$ is depicted. At this temperature, $H_{sp} \sim 0.22$. We find that λ_2 is very small at small H reflecting the coexistence state, and it begins to deviate near H_{SP}. As the size increases, the region where λ_2 is very small approaches H_{SP} as expected. The size dependences of λ_2 are given by the scaling law studied in Sect. 3.4.

5.3.3 *Relaxation times of* 4 × 3 *System*

It is difficult to obtain eigenvalues of short-range interacting systems in a large lattice. In this subsection, however, we dare study the eigenvalues of a model of 4×3 lattice (Fig. 5.10a) to see characteristics of the eigenvalue spectrum of short-range interacting systems.

5.3.3.1 Temperature Dependence of Eigenvalues

The lowest nonzero eigenvalue λ_2 (dotted curve) is quasi-degenerate below $T/J < 2$ which is a symptom of the spontaneous magnetization. The critical temperature in the thermodynamic limit is $T_C \simeq 2.269$.

Some other states also become very small at very low temperatures. This fact indicates that other types of metastable states exist. They come from a metastable domain wall structure, which will be discussed below.

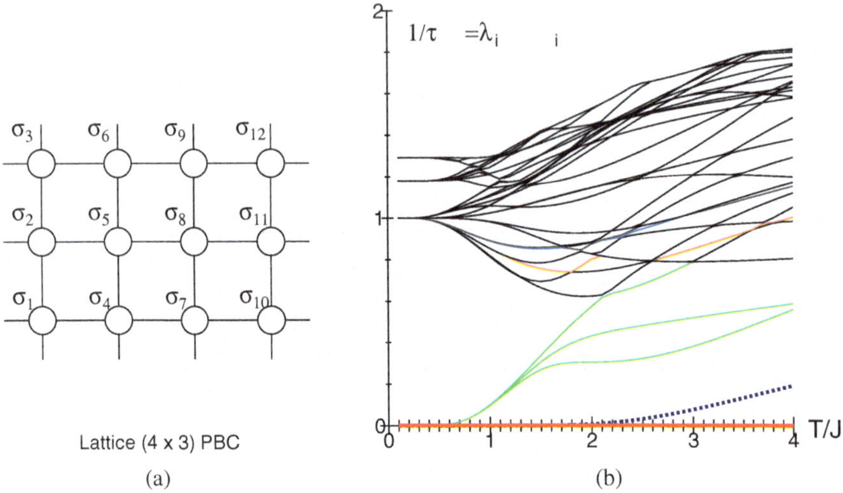

Fig. 5.10 **a** A lattice of 4×3 system with the periodic boundary conditions. **b** Temperature dependence of the eigenvalues of the lattice of 4×3 at $H = 0$. The second eigenvalue is plotted by the dotted curve

Fig. 5.11 **a** Examples of the time evolution of the magnetization at $T/J = 1.8$. **b** That at $T = 3.0$

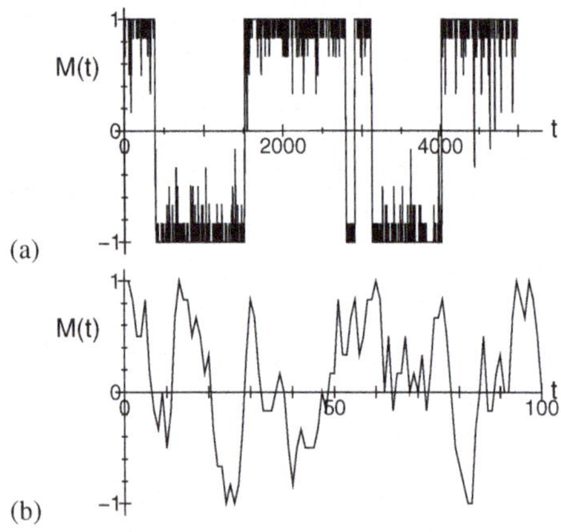

In Fig. 5.11a and b, we depict time evolutions of the magnetization at $T/J = 1.8$ and $T/J = 3$, respectively. The figures are obtained by a Monte Carlo simulation with the transition probability of the master equation. In the Monte Carlo simulation, configurations are updated according to the time evolution of the probability distribution using a random variable. The temperature $T/J = 1.8$ is lower than T_C, and the distribution of the magnetization $P(m)$ has two maxima which correspond to the spontaneous magnetization in the thermodynamic limit.

In a finite lattice, the magnetization stays on one side of $P(m)$ for a duration and jumps to the other side as we find in Fig. 5.11a. The duration corresponds $\tau_2 = 1/\lambda_2$, which depends on the size largely. When the temperature decreases, the lifetime of the plateau increases very rapidly (not shown). The magnetization in a plateau relaxes with $\tau_3 = 1/\lambda_3$.

On the other hand, at $T/J = 3 > T_C$, $P(m)$ has a single maximum at $M = 0$ where fluctuations of the magnetization relax to zero with the relaxation time $\tau_2 = 1/\lambda_2$. In Fig. 5.11b, an example of magnetization dynamics at $T/J = 3$ is depicted, which does not depend much on the size of the system (not shown).

5.3.3.2 Field Dependence of Eigenvalues

In Fig. 5.12, the field dependences of $1/\tau$ at $T/J = 0.1$ and 0.5, which are below the critical temperature $T_C \simeq 2.24$ of the bulk system, are plotted. There, the 50 smallest λ_i $(i = 1, 50)$ are drawn. Many of them are overlapped. The bold dotted line gives λ_2 $(\lambda_1 = 0)$.

At zero temperature, the threshold field for a single spin flip from the perfect ordered state in the square lattice is given by the competition between the gain of energy $-2H$ and the loss of exchange interaction $+8J$, and the threshold is $H = 4J$. Thus, the H dependence of λ_2 is given by

$$\lambda_2 \simeq e^{-\beta\Delta E_{single}} \sim e^{-(8J-2H)/T} \tag{5.71}$$

at a low temperature. At $T/J = 0.1$, we find λ_2 is almost zero degenerating to λ_1 until $H = 4L$. That is, the perfect ordered state has a long relaxation time in this region.

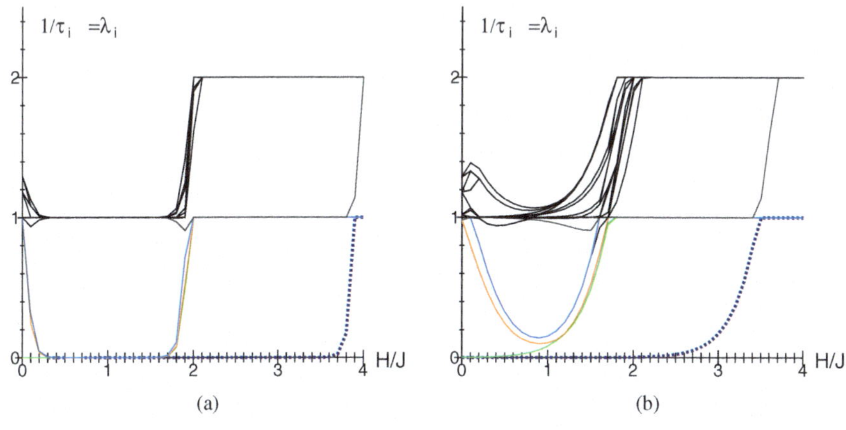

Fig. 5.12 **a** Field dependence of the eigenvalues for a 4×3 system at $T/J = 0.1$. Here, 50 smallest eigenvalues (λ_i $(i = 1, \ldots, 50)$) are drawn. Many of them are overlapped. The bold dotted line gives λ_2 $(\lambda_1 = 0)$. **b** Field dependence of the eigenvalues at $T/J = 0.5$

5.3.3.3 Metastability Due to Flat Domain Wall Structures

In field dependence of eigenvalues (Fig. 5.12), some relaxation of them becomes very small at low temperatures. These indicate that some states have long relaxation times. These slow relaxations come from configurations depicted in Fig. 5.13a. The excitation energy to have a spin flip from the domain wall structure is $4J$. This value is smaller than $8J$ of the single spin flip from the perfect ordered state. Thus, the critical field to have a spin flip from the domain wall is $2H = 4J$, i.e., $H = 2J$, and thus the eigenvalues of these modes begin to increase around $H = 2J$, although λ_2 begins to increase around $H = 4J$.

In Fig. 5.12, we also find a peculiar dependence of eigenvalues, i.e., some of λ_is decrease with H and increase again for a larger H. At $T/J = 0.1$, the eigenvalues become almost 0 around $H/J = 1$. This peculiar behavior is due to special configurations as depicted in Fig. 5.13b. Because the + spin σ_5 at the site (2,2) can flip without energy loss at $H = 0$, this configuration relaxes to other states freely and the corresponding eigenvalues are 1. However, at $H > 0$, the + spin is stabilized due to the Zeeman energy, and the flip of this spin costs $\Delta E = 2H$. Therefore, the relaxation times from this type of configuration become long. This fact causes the bending structure of some eigenvalues. These detailed features of H dependence of the eigenvalues of the kinetic Ising model depend on the structure of the lattice.

The size dependence of the eigenvalues should be noted. Above, we studied the relaxation process in a small lattice and estimated eigenvalues from a viewpoint of a single flip. However, the size dependence of eigenvalues in large systems is not simple. As to λ_2 which is responsible for the symmetry breaking phenomena, it becomes small exponentially with N:

$$\lambda_2 \sim e^{-aN} \tag{5.72}$$

with a positive constant a, while other eigenvalues λ_i coming from metastability of domain walls have weaker dependence, e.g., a power law of N and even constant. Therefore, an essential feature of the metastable state is given by the lowest two eigenvalues λ_2.

Fig. 5.13 **a** A domain wall structure. **b** A stripe structure

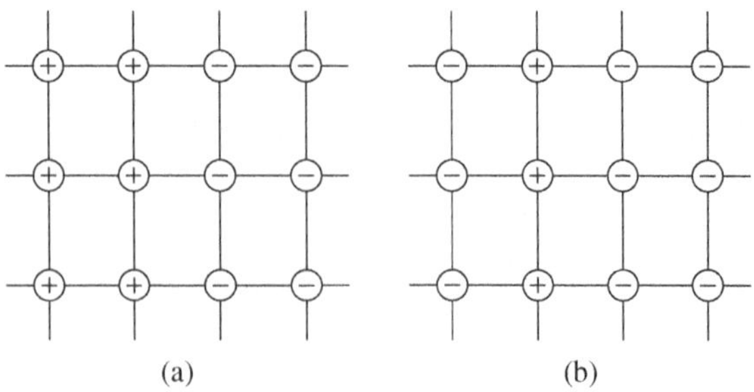

As we saw above, the eigenvalues of the time-evolution operator give some symptoms of metastability, but it is difficult to study enough large systems. To study relaxation time in short-range interacting models, we need some other methods, e.g., estimation free energy landscape $f(m:T,H)$ by Monte Carlo simulation and molecular dynamics by stochastic Landau–Lifshitz–Gilbert equation, which will be explained in Chap. 7.

5.4 Eigenvalue Problem of Quantum Master Equation

The state in a pure quantum system is given by a wave function $|\Psi(t)\rangle$, and it evolves by the Schrödinger equation

$$i\hbar\frac{\partial}{\partial t}|\Psi(t)\rangle = \mathcal{H}|\Psi(t)\rangle, \tag{5.73}$$

where $\mathcal{H}$ is the Hamiltonian of the system. The eigenvalues of the Hamiltonian $\mathcal{H}$ show a peculiar structure for the metastable state, which we will study in Part III.

We may also describe a state of the system by the density matrix ρ

$$\rho(t) = |\Psi(t)\rangle\langle\Psi(t)|. \tag{5.74}$$

The time evolution of the density matrix is given by

$$\begin{aligned}\frac{d}{dt}\rho = \frac{d}{dt}|\Psi(t)\rangle\langle\Psi(t)| &= \left(\frac{1}{i\hbar}\mathcal{H}|\Psi(t)\rangle\right)\langle\Psi(t)| + |\Psi(t)\rangle\left(-\frac{1}{i\hbar}\langle\Psi(t)|\mathcal{H}\right) \\ &= \frac{1}{i\hbar}\left[\mathcal{H},\rho\right].\end{aligned} \tag{5.75}$$

When the system is in a statistical mixture of states, i.e., in the situation in which we only probabilistically know which quantum state really exists and the probability of the state Ψ_k is given by $P(k)$, the density of the state is

$$\rho(t) = \sum_{k=1}^{N} P(k)|\Psi_k(t)\rangle\langle\Psi_k(t)|. \tag{5.76}$$

Here, N is the dimension of the Hamiltonian (not the number of particles) and ρ is an $N \times N$ matrix. In the statistical treatment in dissipative environments, this density matrix for the mixed state is used. Time evolution of the density matrix of the mixed state also follows (5.75).

If we consider the effects of dissipative environments, the time evolution of the system includes a dissipative evolution $\mathcal{D}(\rho)$ which violates the time-reversal symmetry. Then, the time evolution of the density matrix ρ is given by the so-called quantum master equation of the following form:

$$\frac{d}{dt}\rho = \frac{1}{i\hbar}\left[\mathcal{H}, \rho\right] + \mathcal{D}(\rho) \equiv \mathcal{L}\rho. \tag{5.77}$$

Concrete forms of $\mathcal{D}(\rho)$ are given in various ways, e.g., the Lindblad equation and the Redfield equation (see Appendices 8.4.1 and 8.4.2). The time-evolution operator $\mathcal{L}$ is a linear operation on the density matrix, and is called 'super operator'. The number of elements of ρ, M_ρ, is

$$M_\rho = N^2. \tag{5.78}$$

Therefore, when we study the time evolution of ρ as an eigenvalue problem in a standard matrix-vector form, we have to consider an extended space of M_ρ dimensions to express the operation of $\mathcal{L}\rho$. There, we need to introduce a vector $\boldsymbol{v}_\rho$ which is an array of M_ρ elements of ρ. Then, the operation of time evolution is expressed by a matrix $\hat{\mathcal{L}}$ which is a matrix of $M_\rho \times M_\rho$. Then, (5.77) is expressed as

$$\frac{d}{dt}\boldsymbol{v}_\rho = \hat{\mathcal{L}}\boldsymbol{v}_\rho. \tag{5.79}$$

This M_ρ-dimensional space is called 'Hilbert–Schmitt space'.

It should be noted that in classical models, the probability distribution is expressed by $P(i)$ where i is a state of the system. $P(i)$ corresponds to the diagonal element of the density matrix. Therefore, the operation of the time evolution is given by a matrix of $N \times N$ (not $N^2 \times N^2$).

This large dimension of the matrix $\hat{\mathcal{L}}$ prevents us from performing a numerical calculation for systems with many degrees of freedom. In Part III, however, we present an example of the application of this method for a cavity-photon problem.

5.5 Free Energy at the First-Order Phase Transition

We have studied characteristics of metastability by investigating structures of eigenvalues of the transfer matrix and time-evolution operations of systems of many degrees of freedom. The size of matrices rapidly increases with the system size, and it is difficult to study the nature of phase transition explicitly for large systems by this method.

In this section, we discuss a basic picture of the change of state at the first-order phase transition in terms of probability distribution and the free energy function.

If a parameter controlling the stability of the state, e.g., pressure, temperature, and magnetic field, exceeds the transition point, the originally stable state is no more the equilibrium state. But, when the excess of the parameter is small, the original state is locally stable and persists as a metastable state. To make the scenario concrete, we studied the ferromagnetic Ising system below the critical temperature.

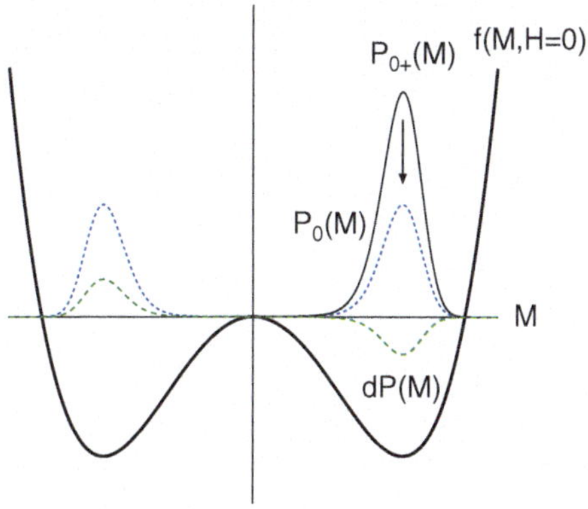

Fig. 5.14 Change of the probability distribution in the positive part, which is realized when the field H is changed from a positive value to zero, $P_{0+}(M)$ (solid line) to the equilibrium distribution $P_{\text{eq}}(M, 0)$ at $H = 0$ (dotted line). The relaxation model $|\phi_2\rangle = \Delta P(M)$ is shown by a dashed line

When the field is positive ($H > 0$), the probability distribution $P(M)$ is located on the positive side of magnetization. Let the distribution be $P_+(M, H)$. For a negative field ($H < 0$), we define $P_-(M, H)$ as well. At the first-order phase transition point $H = 0$, the two states are degenerate, and $P(M)$ has two peaks which are approximately given by

$$P_{\text{eq}}(M, 0) \simeq \frac{P_{0+}(M) + P_{0-}(M)}{2}, \tag{5.80}$$

where

$$P_{0\pm}(M) = \lim_{H \to 0} P_{\pm}(M, H). \tag{5.81}$$

However, when the field changes from a positive H to zero, the probability remains $P(M) = P_{0+}(M, 0)$ representing the spontaneous magnetization. The true equilibrium states is $P_{\text{eq}}(M, 0)$, and thus $P(M)$ begins to relax to $P_{\text{eq}}(M, 0)$ as depicted in Fig. 5.14. There, the relaxation mode $\Delta P(M)$ is shown by a dashed line, which corresponds to $\boldsymbol{P}_{\text{relax}}$ (5.83) given below.

The change of probability is given by a process jumping over the potential barrier between regions of $\pm M$ and it is very small. Kramers' argument (Sect. 3.3) estimates this flow of probability in an approximate way. This picture may be expressed by a simplified two-level system with the two states S_+ and S_- which are the states with positive and negative magnetizations, respectively. The time evolution is given by

$$\frac{d}{dt}\begin{pmatrix} P_+(t) \\ P_-(t) \end{pmatrix} = \begin{pmatrix} -r & r \\ r & -r \end{pmatrix}\begin{pmatrix} P_+(t) \\ P_-(t) \end{pmatrix}, \tag{5.82}$$

where r is the rate of change between $\pm M$ states per unit time. The eigenstates for the equilibrium and that of relaxation are given by

$$\boldsymbol{P}_{\text{eq}} = \frac{1}{2}\begin{pmatrix} 1 \\ 1 \end{pmatrix}, \quad \boldsymbol{P}_{\text{relax}} = \frac{1}{2}\begin{pmatrix} 1 \\ -1 \end{pmatrix}, \tag{5.83}$$

respectively. The eigenvalues are

$$\hat{\lambda}_{\mathrm{eq}} = 0, \quad \hat{\lambda}_{\mathrm{relax}} = -2r, \tag{5.84}$$

respectively. The initial state

$$\boldsymbol{P}_0 = \begin{pmatrix} 1 \\ 0 \end{pmatrix} = \boldsymbol{P}_{\mathrm{eq}} + \boldsymbol{P}_{\mathrm{relax}} \tag{5.85}$$

changes as

$$\boldsymbol{P}(t) = \boldsymbol{P}_{\mathrm{eq}} + e^{-2rt}\boldsymbol{P}_{\mathrm{relax}}. \tag{5.86}$$

If we consider this situation from a viewpoint of free energy, we have the following image. For $H > 0$, the free energy function of the magnetization is given by

$$F_+(M, H) = -k_{\mathrm{B}}T \ln P_+(M, H) \tag{5.87}$$

which is located in the positive magnetization region, and the contribution from the negative M region can be ignored in a large system. For $H < 0$, similarly we have $F_-(M, H)$. However, when $H = 0$, the situation is subtle, and we have to take both sides into account. Thus, the free energy has a singularity at $H = 0$.

This singularity appears as a problem of convergence of the high-field expansion of the free energy:

$$f_+(T, H) = \sum_{n=0}^{\infty} c_{m,n} y^m z^n, \quad y = e^{-2\beta J}, z = e^{-2\beta H} \tag{5.88}$$

at $H = 0$ below the critical temperature ($T < T_{\mathrm{C}}$). For a positive field, the series converges, but when the field changes from a positive value to a negative value, the state does not change much because of the metastability, and apparent free energy for the state in the negative H may be expressed by an analytic continuation of $f_+(T, H)$ to the negative field ($H > 0 \to H < 0$). In this case, the series becomes an asymptotic expansion, and the sum does not converge.

Langer [2, 3] gave detailed analyses on this problem, which will be briefly reviewed in the next section.

5.6 Langer's Argument

Langer gave essential viewpoints for singular properties of the free energy for the metastable state due to the first-order phase transition [2, 3]. He gave three types of arguments, which clarified important features of metastability.

5.6.1 Langer's Analysis I: A Picture of Nucleation Cluster

The first one is an argument for the asymptotic property of contributions from large droplets (anti-magnetized domains) to a kind of high-field expansion of the free energy. He considered a droplet model in a d-dimensional system. A large droplet of l antiparallel spins has the energy:

$$\phi_l \simeq 2Hl + \sigma l^{(d-1)/d}, \tag{5.89}$$

as discussed in Sect. 3.5. Here, the first term is the Zeeman energy and the second one is the surface energy of the droplet where σ is the interface energy par a unit area.

In the symmetry-broken state, the partition function is expanded in a high-field expansion style:

$$Z(T, H) = e^{N\beta H - \beta N E_0}\left(1 + \sum_{l=1}^{\infty} N \exp\left(-\beta(2Hl + \sigma l^{(d-1)/d})\right)\right), \tag{5.90}$$

where NE_0 is the ground state energy, and the free energy is given by

$$\begin{aligned} F(T, H) = -k_B T \ln Z(T, H) &\simeq NE_0 - NH - k_B T \ln\left(1 + \sum_{l=1}^{\infty} N \exp\left(-\beta(2Hl + \sigma l^{(d-1)/d})\right)\right) \\ &\simeq NE_0 - NH - Nk_B T \sum_{l=1}^{\infty} \exp\left(-\beta(2Hl + \sigma l^{(d-1)/d})\right). \end{aligned} \tag{5.91}$$

The magnetization is given by

$$M = -\left(\frac{\partial F}{\partial H}\right) = N + \frac{N}{\beta}\frac{\partial}{\partial H}\left(\sum_{l=1}^{\infty} \exp\left(-\beta(2Hl + \sigma l^{(d-1)/d})\right)\right). \tag{5.92}$$

Now, we define a function:

$$f(H) = \sum_{l=1}^{\infty} \exp\left(-\beta(2Hl + \sigma l^{(d-1)/d})\right). \tag{5.93}$$

In the negative field, this sum diverges because the term $2Hl + \sigma l^{(d-1)/d}$ becomes negative for large l. As for the free energy for the metastable state, the sum of l should be cut off at

$$l_C = \left(\frac{(d-1)\sigma}{2d|H|}\right)^d, \quad d \geq 2, \tag{5.94}$$

until which the energy increases with l. Here, l_C corresponds to the critical size of nucleation. Thus, the sum is ill-defined for negative H. To characterize the metastability, Langer introduced an analytic continuation of $f(H)$.

Let the dimension d be three for concreteness, and using a variable t defined as

$$l = \frac{\sigma^2}{8H^3} t^3. \tag{5.95}$$

Then, $f(H)$ is rewritten as

$$f(H) = \frac{3\sigma^2}{8H^3} \int_0^\infty t^2 \exp\left(-\frac{\beta\sigma^3}{4H^3}(t^3 + t^2)\right) dt. \tag{5.96}$$

The summation of l for a positive H corresponds to the integral along the contour $t = 0$ to ∞, which has no singularity because the system is stable for $H > 0$. For a negative H, the summation has a discrepancy as we see above. This situation corresponds the integral along the contour from $t = 0$ to negative direction of t. The discrepancy is analyzed by making use of an analytic continuation of $f(H)$ in the complex t plane from positive H to $-H$.

In the complex t plane, the original integration contour C_1 $(0, \infty)$ along the real t axis is transferred to the integration contour C_2 for $H \to He^{i\pi}$ or C_3 for $H \to He^{-i\pi}$. The contours must avoid the region of $\mathrm{Re}(t^3 + t^2) < 0$. In these transformations, $f(H)$ has a discontinuity which is given by the difference of the integrals along C_2 and C_3. Because $|H|$ is small, we may estimate the integrals by the saddle point method.

Let $t = x + iy$, and then the real part of $t^3 + t^2$ be

$$\mathrm{Re}(t^3 + t^2) = x^3 - 3xy^2 + x^2 - y^2. \tag{5.97}$$

The saddle points are given the relations:

$$\frac{d\mathrm{Re}(t^3 + t^2)}{dx} = 3x^2 - 3y^2 + 2x = 0, \quad \frac{d\mathrm{Re}(t^3 + t^2)}{dy} = -6xy - 2 = 0. \tag{5.98}$$

They are at

$$t = 0, \quad t = -\frac{2}{3}. \tag{5.99}$$

The contours C_2 and C_3 are taken along the extreme points in the complex t plane. And the difference is given by a contour $C_4 = C_2 - C_3$ which goes through the point $t = -\frac{2}{3}$ along the imaginary direction. By estimating the integral along C_4 by a saddle point method at $t = -2/3$, the difference of the free energy is given by

$$f(H_2) - f(H_3) = \frac{3\sigma^2}{8H^3} \times \sqrt{-\frac{\pi H^2}{\beta\sigma^3}} \times t_{\text{saddle}}^2 \exp\left(-\frac{\beta\sigma^3}{4H^2}(t_{\text{saddle}}^3 + t_{\text{saddle}}^2)\right)$$

$$= -i\frac{\sigma^{3/2}\pi^{1/2}}{3\beta^{1/2}H^2}\exp\left(-\frac{\beta\sigma^3}{27H^2}\right). \quad (5.100)$$

Thus,

$$f(|H| \pm i0) = \bar{f}(H) \pm i\Delta(H), \quad (5.101)$$

where

$$\bar{f}(H) = \frac{3\sigma^2}{8H^3}\int_0^{-2/3} t^2 \exp\left(-\frac{\beta\sigma^3}{4H^2}(t^3 + t^2)\right) dt \quad (5.102)$$

which gives apparent free energy of the metastable state and the difference was obtained as

$$\Delta(H) \simeq \frac{f(H_2) - f(H_3)}{2} = -i\frac{\sigma^{3/2}\sqrt{\pi}}{3\sqrt{\beta}H^2}\exp\left(-\frac{\beta\sigma^2}{27H^2}\right). \quad (5.103)$$

In this way, the metastability is characterized by an appearance of imaginary part in the free energy.

5.6.2 *Langer's Analysis II: Functional Integral*

Langer also studied the singularity of the partition function conceptually in a functional integral of a ϕ^4 type model.

In the transfer matrix method, we considered the partition function specifying the edge states $(\sigma_{11}, \sigma_{12}, \ldots \sigma_{1M})$ and $(\sigma_{N1}, \sigma_{N2}, \ldots \sigma_{NM})$. There we integrate out all the internal spins $(\sigma_{i1}, \sigma_{i2}, \ldots \sigma_{iM})$, $i = 2, \ldots N-1$. This idea is the same as that of the path-integral method in which we consider the sum of weights of all the possible paths specifying the initial and final states.

For a one-dimensional array of magnetization $\{m_i\}, i = 1, N$, instead of the spin configuration, the partition function is given by

$$Z = \prod_{i=1}^{N}\left(\int_{-\infty}^{\infty} \frac{dm_i}{\sqrt{2\pi a}}\right)\exp\left(-a\sum_{i=1}^{N}\left[\frac{(m_{i+1} - m_i)^2}{2a^2} + \epsilon m_i^2 + \alpha m_i^4\right]\right), \quad (5.104)$$

where a is the distance between the neighboring sites. In a continuum limit, $N \to \infty$, $a \to 0$ keeping $Na = L$ finite. The partition function is given by

$$Z(\alpha) = \int \delta m(x) \exp\left(- \int_0^L dx \left[\frac{1}{2} \left(\frac{dm}{dx} \right)^2 + \epsilon m^2 + \alpha m^4 \right] \right), \tag{5.105}$$

where $m(x)$ is the spatial profile of the order parameter, e.g., the magnetization. The integral of $\delta m(x)$ denotes the summation of all the paths of $m(x)$ which is defined as

$$\int \delta m(x) = \lim_{N\to\infty} \prod_{j=1}^{N} \left(\int_{-\infty}^{\infty} \frac{dm_j}{\sqrt{2\pi L/N}} \right), \tag{5.106}$$

where $m_j = m(x = jL/n)$, and the integral of $\delta m(x)$ denotes summation of all the paths of $m(x)$ fixing

$$m(0) = m_0, \quad m(L) = m_L. \tag{5.107}$$

The integral converges for positive α but diverges for negative α. Thus, $Z(\alpha)$ has a singularity at $\alpha = 0$. Langer showed that the α dependence of the partition function has an essential singularity which is similar to that we saw above.

The partition function (5.105) is regarded as an integral over the Wiener measure [55, 56] and can be analyzed by a spectrum representation of the Green function given by

$$G(m_0, m_L, L) = \int_{m_0}^{m_L} \delta m(x) \exp\left(- \int_0^L dx \left[\frac{1}{2} \left(\frac{dm}{dx} \right)^2 + \epsilon m^2 + \alpha m^4 \right] \right). \tag{5.108}$$

It is known that the Green function is given by a spectral representation of the form

$$G(m_0, m_L, L) = \sum_{n=0}^{\infty} \psi_n(m_0)\psi(m_L)e^{-\lambda_n L}, \tag{5.109}$$

where $\psi_n(m)$ and λ_n are the nth eigenfunction and eigenvalue of the Schrödinger equation [57, 58]:

$$\left[-\frac{1}{2}\frac{d^2}{dx^2} + (\epsilon m^2 + \alpha m^4) \right] \psi_n = \lambda_n \psi_n. \tag{5.110}$$

The partition function is given by

$$Z(\alpha) = \int_{-\infty}^{\infty} dm G(m, m, L) = \sum_{n=0}^{\infty} e^{-\lambda_n L} \to e^{-\lambda_0 L}. \tag{5.111}$$

Thus, in the limit $L \to \infty$, the free energy per unit length is

$$f(\alpha) = -k_B T \lim_{L\to\infty} \frac{1}{L} \ln Z(\alpha). \quad (5.112)$$

So, we study the ground state eigenvalue of the Schrödinger equation (5.110). For a positive α, a normalizable wave function exists, while for a negative α the potential function diverges negatively, and no normalizable wave function exists. Thus, $\alpha = 0$ is a singular point. But we still try to obtain λ_0 by an analytic continuation. In the case $\alpha < 0$, there is an outgoing current, and such a boundary condition causes the Hamiltonian to be non-Hermitian and the eigenvalue becomes complex. These types of phenomena are extensively studied as the non-Hermite phenomena [59]. To simplify the equation, let us use new variables:

$$m = (2\epsilon)^{-1/4}\xi, \quad \gamma = \frac{1}{\sqrt{2}}\epsilon^{-3/2}|\alpha|, \quad \eta = 1 - \sqrt{2}\epsilon^{-1/2}\lambda_0. \quad (5.113)$$

Then, the Schrödinger equation for $\phi(\xi) = \psi_0(m)$ is given by

$$\left[-\frac{d^2}{d\xi^2} + (\xi^2 - \gamma\xi^4)\right]\phi = (1-\eta)\phi. \quad (5.114)$$

Instead of obtaining the solution, Langer estimated the imaginary part of the eigenvalue using the relation:

$$\mathrm{Im}(\eta) = \frac{J(\xi)}{\int_0^\xi |\phi(\xi')|^2 d\xi'}, \quad (5.115)$$

where $J(\xi)$ is the current given by

$$J(\xi) = \frac{1}{2i}\left(\phi^*(\xi)\frac{d\phi}{d\xi} - \frac{d\phi^*}{d\xi}\phi(\xi)\right). \quad (5.116)$$

This quantity can be estimated from the wave function obtained by the WKB approximation. Outside of the turning point, the outgoing wave is

$$\phi_{\mathrm{WKB}} \simeq \frac{1}{\sqrt{p}} \exp\left(\pm i \int_{\xi_{\mathrm{turning}}}^{\xi} p(\xi')d\xi' \pm \frac{\pi}{4}\right), \quad (5.117)$$

where

$$p(\xi) = \sqrt{E - V(\xi)} = \sqrt{1 - \eta - \xi^2 - \gamma\xi^4}. \quad (5.118)$$

From this form of ϕ_{WKB}, the current is obtained as

$$J(\xi) = \pm p|\phi_{\mathrm{WKB}}|^2 = \pm 1. \quad (5.119)$$

When α is small, the wave function inside the barrier is almost to be the ground state wave function of the harmonic oscillator:

$$\phi(x) = Ae^{-\frac{1}{2}\xi^2}, \tag{5.120}$$

where the coefficient A is determined by the wave function obtained by the connection formula of the WKB method and (5.120), which gives

$$A \simeq \frac{\gamma^{1/4}}{\sqrt{2}} e^{\frac{1}{3\gamma}}. \tag{5.121}$$

Now, we obtain (5.115) as

$$\mathrm{Im}(\eta) = \pm\frac{2}{\sqrt{\pi}}A^2 = \pm\frac{4}{\sqrt{\pi\gamma}}e^{-\frac{2}{3\gamma}}. \tag{5.122}$$

The imaginary part of the free energy is

$$\mathrm{Im}(f(\alpha)) = k_{\mathrm{B}}T\,\mathrm{Im}(\lambda_0) \simeq \pm k_{\mathrm{B}}T\frac{2^{7/4}\epsilon^{5/4}}{\sqrt{\pi}|\alpha|}\exp\left(-\frac{(2\epsilon)^{3/2}}{3|\alpha|}\right). \tag{5.123}$$

This argument gives a fancy picture of the appearance of a metastable state.

5.6.3 Langer's Analysis III: A Picture of the Action

Langer also estimated the partition function (5.105) by the saddle point method. Let the exponent in the functional integral (action) be

$$S(m(x)) = \int_0^L dx\left[\frac{1}{2}\left(\frac{dm}{dx}\right)^2 + \epsilon m^2 + \alpha m^4\right]. \tag{5.124}$$

The minimum of S gives the main contribution to the partition function. In the analogy to terminologies in the analytical mechanics, regarding x to be the time and $m(x)$ to be the particle position, we may regard the integrand as the Lagrangian. The corresponding Euler–Lagrangian equation is

$$-\frac{1}{2}\frac{d^2m}{dx^2} + \epsilon m + 2\alpha m^3 = 0. \tag{5.125}$$

For positive α, the only bounded solution is $m(x) = 0$, which makes S be minimum. In the case $\alpha < 0$, (5.125) has nonzero solutions, which give the motion of the particle in the potential

$$V(x) = -\epsilon m^2 + |\alpha| m^4. \tag{5.126}$$

Noting the relation:

$$\frac{d}{dx}\text{sech}(x) = -\frac{\sinh(x)}{\cosh^2(x)}, \quad \frac{d^2}{dx^2}\text{sech}(x) = \text{sech}(x) - 2\text{sech}^3(x), \tag{5.127}$$

the bounded solution of (5.125) is given by

$$m_{\text{bs}}(x) = \pm\sqrt{\frac{\epsilon}{|\alpha|}}\text{sech}\left(\sqrt{2\epsilon}(x - x_0)\right). \tag{5.128}$$

When $L \to \infty$, we may take arbitrary value of x_0 except those very close to 0 or L. For this solution, the partition function (5.105) is given by

$$\exp\left(-S(m_{\text{bs}}(x))\right) = \exp\left(-\frac{(2\epsilon)^{3/2}}{3|\alpha|}\right). \tag{5.129}$$

It should be noted that $S(m_{\text{bs}}(x))$ does not give the maximum of $\exp(-S(x))$. In some direction around m_{bs}, $S(m_{\text{bs}}(x))$ decreases reflecting the metastability. To see this property, we expand $m(x)$ around m_{bs}:

$$m(x) = m_{\text{bs}}(x) + \nu(x). \tag{5.130}$$

Up to the second order of $\nu(x)$, the action is given by

$$S(m(x)) \simeq S(m_{\text{bs}}(x)) + \int_0^L dx \left(\frac{1}{2}\left(\frac{d\nu}{dx}\right)^2 + \epsilon\nu^2 - 6|\alpha| m_{\text{bs}}(x)^2\nu^2\right). \tag{5.131}$$

To be a solution of (5.125), $\nu(x)$ satisfies the following eigenvalue equation:

$$-\frac{1}{2}\frac{d^2\nu}{dx^2}\nu_k(x) + \epsilon\nu - 6|\alpha| m_{\text{bs}}(x)\nu_k(x) = 0. \tag{5.132}$$

Using the solution of the following Schrödinger equation :

$$-\frac{1}{2}\frac{d^2\nu}{dx^2}\nu_k(x) - 6|\alpha| m_{\text{bs}}(x)\nu_k(x) = E_k\nu_k(x), \tag{5.133}$$

we express a small deviation from $m_{\text{bs}}(x)$ as

$$\nu(x) = \sum_k \zeta_k \nu_k(x). \tag{5.134}$$

Then, $S(m(x))$ is given by

$$S(m(x)) = S(m_{\mathrm{bs}}(x)) + \sum_k (\epsilon + E_k)\zeta_k^2. \tag{5.135}$$

Using the explicit form of $m_{\mathrm{bs}}(x)$ (5.128), the Schrödinger equation is given by

$$-\frac{1}{2}\frac{d^1}{dx^2}\nu_k(x) - 6\epsilon\,\mathrm{sech}^2(\sqrt{2\epsilon}x) = E_k\nu_k(x), \quad k = 0, 1, \ldots. \tag{5.136}$$

This equation has two eigenstates with negative eigenvalues [60]:

$$\nu_0(x) = \sqrt{\frac{3}{4}}(2\epsilon)^{1/4}\mathrm{sech}^2(\sqrt{2\epsilon}x), \quad E_0 = -4\epsilon \tag{5.137}$$

and

$$\nu_1(x) = \sqrt{\frac{3}{2}}(2\epsilon)^{1/4}\frac{\sinh(\sqrt{2\epsilon}x)}{\cosh^2(\sqrt{2\epsilon}x)}, \quad E_1 = -\epsilon. \tag{5.138}$$

The modes ν_k give contributions to $S(m(x))$ as

$$S(m(x)) \simeq S(m_{\mathrm{bs}}(x)) + \sum_{k=0}(\epsilon + E_k)\zeta_k^2 = S(m_{\mathrm{bs}}(x)) - 3\epsilon\zeta_0^2 + 0\times\zeta_1^2 + \sum_{k\geq 2}(\epsilon + E_k)\zeta_k^2. \tag{5.139}$$

For $k \geq 2$, $\epsilon + E_k > 0$ and the integrals for the variables $\zeta_k (k \geq 2)$ converge. But the first mode ζ_0 causes the integral to diverge.

In this situation, we again consider the analytic continuation from positive to negative α. Let the partition function for $\alpha = 0$ be Z_0, and we may express the partition function as

$$Z \sim Z_0 + Z_1(\alpha). \tag{5.140}$$

The contribution due to negative α is given by

$$Z_1(\alpha) \simeq e^{-S(m_{\mathrm{bs}}(x))}\int d\zeta_0 e^{3\epsilon\zeta_0^2} \times \int d\zeta_1 e^{0\times\epsilon\zeta_1^2} \times \int d\zeta_2 e^{-(\epsilon+E_2)\zeta_2^2} \times \cdots. \tag{5.141}$$

Here, $e^{-S(m_{\mathrm{bs}}(x))}$ is given by (5.129) and the integral for ζ_0 in the analytic continuation gives

$$e^{-S(m_{\mathrm{bs}}(x))}\int_0^{\pm i\infty} d\zeta_0 e^{3\epsilon\zeta_0^2} = \pm\frac{i}{2}\sqrt{\frac{\pi}{3\epsilon}} \times \exp\left(-\frac{(2\epsilon)^{3/2}}{3|\alpha|}\right). \tag{5.142}$$

The integral for ζ_1 gives a contribution proportional to L:

$$2L\frac{(2\epsilon)^{3/4}}{(3|\alpha|)^{1/2}}. \tag{5.143}$$

The kth integrals ($k \geq 2$) gives

$$\sqrt{\frac{\pi}{\epsilon + E_k}}. \tag{5.144}$$

Taking into account the contributions, we have

$$Z_1 = \pm i L A_0 \sqrt{\frac{\pi}{3\epsilon}} \exp\left(-\frac{(2\epsilon)^{3/2}}{3|\alpha|}\right) \frac{(2\epsilon)^{3/4}}{(3|\alpha|)^{1/2}} \prod_{k=2}^{\infty} \sqrt{\frac{\pi}{\epsilon + E_k}}, \tag{5.145}$$

where A_0 is a normalization constant coming from the factor $1/\sqrt{2\pi a}$ in (5.104). For small Z_1, the partition function is given by

$$Z = Z_0(\alpha) + Z_1(\alpha) = Z_0\left(1 + \frac{Z_1}{Z_0}\right) \simeq Z_0 \exp\left(\frac{Z_1}{Z_0}\right). \tag{5.146}$$

The imaginary part of the free energy is estimated as

$$\mathrm{Im}\frac{f(\alpha)}{k_B T} = -\mathrm{Im}\frac{1}{L} Z(\alpha) = -\frac{1}{iL}\frac{Z_1}{Z_0}. \tag{5.147}$$

By estimating the contribution from the integrals for $k = 1, 2, \ldots$ and from Z_0, $\mathrm{Im} f(\alpha)$ was shown to have the same value as (5.123) [2].

5.6.4 Langer's Estimation of Decay Rate of Metastable State

In the previous subsections, we have seen three types of arguments, and found the same results in them. The arguments bring us a mathematical scheme and a conceptual understanding of metastability. And this type of treatment has been repeatedly done in the literature.

Langer also studied the relationship of the imaginary part of the free energy and the decay rate of metastable state which is suggested by the relation (5.115) [3]. The current from the metastable state to the stable state is given by

$$I = |\kappa| \left(\frac{k_B T}{2\pi |\lambda_1|}\right)^{1/2} V \frac{e^{-\bar{E}/k_B T}}{Z_0} \prod_{k:\lambda_k > 0} \left(\frac{k_B T}{2\pi |\lambda_k|}\right)^{1/2}, \tag{5.148}$$

where κ/λ_k is the eigenvalue of a matrix $\tilde{M}$ for the current, and Z_0 is the partition function of the metastable phase. By comparing the relation (5.147), the following result is obtained.

$$I = \frac{|\kappa|}{\pi k_B T} \mathrm{Im}\mathcal{F}, \tag{5.149}$$

where $\mathcal{F}$ is the free energy of the system obtained by the analytic continuation.

Part III
Metastability in Quantum Systems

Chapter 6
Collapse of Metastability by the Quantum Fluctuation

6.1 Introduction

In this chapter, we study metastability in quantum systems. Relaxation of metastable state by quantum fluctuation has been studied as quantum tunneling. Quantum tunneling through the energy barrier at a low temperature has attracted interests [6, 7, 61, 62].

At finite temperatures, thermal fluctuation causes the relaxation of metastable state, where the relaxation time of metastable state due to thermal agitation is given by the Arrhenius law and it increases exponentially with the temperature. In contrast, the relaxation due to the quantum tunneling does not depend on the temperature and the relaxation time saturates at low temperatures and it is finite even at zero temperature. Thus, the breakdown of metastable state due to the quantum tunneling is a practical problem in the downsizing of devices. When the size of a system becomes small, relaxation time of metastability becomes short and the memory in the devise would be erased out.

This problem has been discussed as a problem of macroscopic quantum tunneling. Quantum tunneling effects would cause various interesting phenomena in a macroscopic scale, although it should be noted that, even in the cases where thermal fluctuation causes the relaxation, saturation of relaxation time at low temperature is also observed due to a distribution of system sizes [63].

In pure quantum mechanical motion, we study the dynamics of wave function by Schrödinger equation, where the energy is conserved. Thus, the idea of 'metastable state' is not well defined. However, we may consider a case in which a Hamiltonian has a double-well potential and the initial wave function is localized in one of the wells. The wave function evolves in time and some population transfers to the other well. We call the transition of the population 'quantum tunneling' which corresponds to decay of metastable state.

In quantum systems, operations of the momentum and position do not commute, and structure of eigenstates of the Hamiltonian contains information of dynamical

S. Miyashita, *Collapse of Metastability*, Fundamental Theories of Physics 211,
https://doi.org/10.1007/978-981-19-6668-2_6

nature of the system. Thus, in this section, metastability of pure quantum systems is studied from a viewpoint of eigenvalue structure of the system Hamiltonian.

The second-order phase transition at $T = 0$ is associated with a condensation of the energy levels at which order parameter changes critically but continuously. For example, the order-disorder phase transition of the transverse Ising model (see Sect. 6.12.1) and the dimer-Haldane phase transition [64] are the typical example. In the present chapter, we study the quantum first-order phase transition in many degrees of freedom from a viewpoint of eigenvalue structure of the Hamiltonian and find a characteristic structure for the first-order phase transition.

Here, we note on the first-order quantum phase transition in the ground state. At $T = 0$, the energy and the free energy are the same, and they change continuously with parameters. Thus, the concept of latent heat does not exist. If energy levels of the two states $|\mathrm{A}\rangle$ and $|\mathrm{B}\rangle$ simply cross when a parameter of the Hamiltonian is swept, the expectation value of a quantity O in the ground state changes discontinuously from $\langle \mathrm{A}|O|\mathrm{A}\rangle$ to $\langle \mathrm{B}|O|\mathrm{B}\rangle$, which is generally discontinuous. Thus, the simple cross of energy levels may be regarded as a first-order phase transition. But it is not a phase transition of the type in systems of many degrees of freedom. In the present chapter, we study systems in which a metastable state is separated from the stable state and between them many intermediate states with higher energies exist.

We explore of properties of metastability in several quantum systems. First, we study motion of a particle in a double-well type potential in Sect. 6.2. Then, as a problem of particle conveyance [65], we study a quantum mechanical escape of a particle from a penitential-well when the well is swept (Sect. 6.4). There, we study the relaxation of initial state in metastable potential, which shows an exponential decay in time although the dynamics is purely quantum mechanical. This type of problem has been extensively studied as the resonance state in the scattering problem [30, 59, 66–69, 74]. The mechanism of the exponential relaxation is discussed.

Quantum tunneling is also an important subject in quantum magnets. In Sect. 6.5, dynamics of magnetization in single-molecular magnets in a sweeping field is studied, where we study the effects of environments. As an extended model ($S \to \infty$), the quantum Stoner–Wohlfarth model is also studied in Sect. 6.10. The effect of environment on the relaxation has been studied extensively [75]. In the study, a model with lowest two levels representing the hybridized two stable states are usually concerned. There, only the diagonal component of noise plays an important role as Caldeira–Leggett [75, 76] pointed out, which is also discussed.

As to the quantum phase transition, the transverse Ising model is studied in Sect. 6.12. There, the nature of quantum fluctuation and thermal fluctuation are compared by making use of configurations obtained by a quantum Monte Carlo simulation. We also study the first-order phase transition in dynamical systems. As an example, mechanism of the optical bistability is studied in a cavity system consisting of photon and spins in a cavity in Sect. 6.14. A discontinuous change of transmission as a function of strength of input light is found. Size dependence of the relaxation time is studied, and also a limit cycle under an oscillation external field is investigated.

Besides above-mentioned topics, the tunneling phenomenon is also applied to a variety of systems, e.g., depinning process of trapped domain wall at a potential well due to crystal defect or impurities [6], which have common features with the topics studied in this chapter. The tunneling effect for magnetic domain pinning has been also discussed [62]. As to the quantum tunneling, or relaxation of metastable state, various works have been done. For example, Affleck [5] discussed quantum-statistical metastability. Callen and Coleman studied the metastability in quantum systems as a problem of fate of the false vacuum [4], which give conceptual ideas of the relaxation of metastable. For the details, please find the original works and appropriate reviews [7].

6.2 Quantum Mechanical States in Double-Well Type Potential

Let us begin with a simple model of metastability. We consider a system of a particle in a potential with two local minima which we call 'double-well potential'. Let us consider two trapping potentials $V_1(x)$ and $V_2(x)$ and let them be hybridized by an exchange Γ. We construct the total system in the following way. At each position x, the particle is trapped in the potential $V_1(x)$ or $V_2(x)$ which are coupled by Γ. This situation is expressed by the following matrix:

$$\begin{pmatrix} \mathcal{H}_1(x) & \Gamma \\ \Gamma & \mathcal{H}_2(x) \end{pmatrix}. \tag{6.1}$$

The matrix element for the hybridization Γ may depend on x, but here we adopt a constant Γ for the simplicity. We use the eigenvalue of this system

$$V_{\text{hybrid}\pm}(x) = \frac{V_1(x) + V_2(x) \pm \sqrt{(V_1(x) - V_2(x))^2 + 4\Gamma^2}}{2} \tag{6.2}$$

for the potential. Here, we consider potentials $V_1(x)$, $V_2(x)$:

$$V_1(x) = \frac{1}{2}\left(x + \frac{1}{2}\right)^2, \quad V_2(x) = \frac{1}{2}\left(x - \frac{1}{2}\right)^2 \tag{6.3}$$

which have minimum at $x = -1/2$ and $x = 1/2$, respectively. The hybridized potentials are depicted in Fig. 6.1a. If we add an external field H,

$$V_1(x) = \frac{1}{2}\left(x + \frac{1}{2}\right)^2 - Hx, \quad V_2(x) = \frac{1}{2}\left(x - \frac{1}{2}\right)^2 - Hx \tag{6.4}$$

the potential is tilted as shown in Fig. 6.1b.

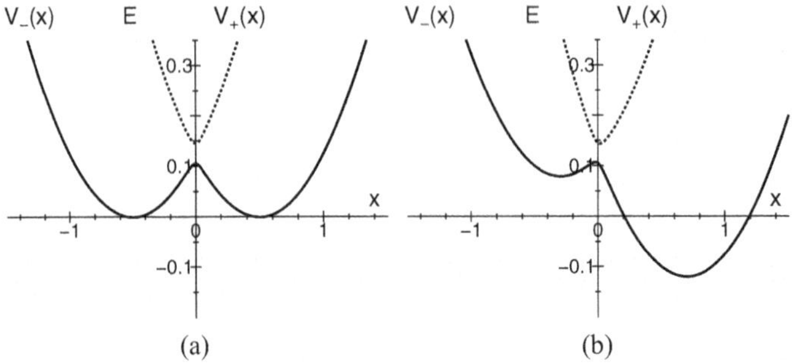

Fig. 6.1 **a** Hybrid potential for $\frac{1}{2}\left(x+\frac{1}{2}\right)^2 - Hx$ and $V_2(x) = \frac{1}{2}\left(x-\frac{1}{2}\right)^2 - Hx$. The exchange constant $\Gamma = 0.02$ and H=0. The solid line denotes $V_{\mathrm{hybrid}-}$, and the dotted line denotes $V_{\mathrm{hybrid}+}$. **b** $H = 0.2$

Here, we adopt the lower branch $V_{\mathrm{hybrid}-}(x)$ as a potential function,[1]

$$V(x) = V_{\mathrm{hybrid}-}(x), \tag{6.5}$$

and study a wave function of the system with this potential.

When $H = 0$, $V(x)$ has a double-well shape, and when $H = 0.2$, the potential is tilted and the left minimum becomes metastable. The Schrödinger equation is given by

$$\left(-\frac{\hbar^2}{2m}\frac{d^2}{dx^2} + V(x)\right)\psi(x) = E\psi(x). \tag{6.6}$$

We solve it by a numerical way and obtain eigenvalues $\{E_i\}$ and the eigenfunctions $\{\psi(x)_i\}$, which are depicted in Fig. 6.2.

In the case of $H = 0$ (Fig. 6.2a), the ground states in both wells are hybridized and form a bonding state (the lowest energy) and an antibonding state (the second lowest energy) with a small energy gap. The coefficient $\hbar^2/m$ is a parameter to control the strength of quantum fluctuation. In Fig. 6.2a and b, we took

$$\frac{\hbar^2}{2m} = 0.005. \tag{6.7}$$

In the present potential, the eigenvalues of the system with the potentials $V_1(x)$ and $V_2(x)$ are the same. The eigenvalues of each potential are given by

$$E_n^{(0)} = \left(\frac{1}{2} + n\right)\hbar\omega, \quad n = 0, 1, \ldots. \tag{6.8}$$

[1] $V_{\mathrm{hybrid}+}(x)$ gives a higher branch plotted by the dotted line. This branch may be regarded as a quasi-potential function of the metastable state, and it can be important when the system is strongly excited, but here we do not consider it.

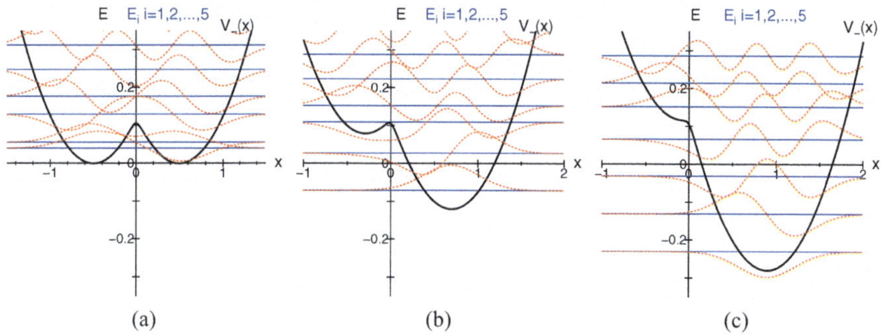

Fig. 6.2 **a** Eigenvalues (horizontal blue line) and eigenstates (red dotted line) for the hybrid potential for $V(x) = (V_1(x) + V_2(x) \pm \sqrt{(V_1(x) - V_2(x))^2 + 4\Gamma^2})/2$ plotted by a bold solid line. Here, $V_1(x)$ and $V_2(x)$ are given in (6.4) with $\Gamma = 0.02$, and $H = 0$, **b** $H = 0.2$, **c** $H = 0.4$. $\frac{\hbar^2}{2m} = 0.005$

In the calculation, we took the value of k to be 1, and thus

$$\frac{\hbar^2 k}{m} = (\hbar\omega)^2 = 0.01 \tag{6.9}$$

and the separation of eigenvalues is

$$\hbar\omega = 0.1, \tag{6.10}$$

and the ground state energy is

$$E_0^{(0)} = 0.05. \tag{6.11}$$

After the hybridization due to Γ, the two ground states form bonding and antibonding states and the eigenvalues are given by

$$E_\pm(0) = E_1^{(0)} \pm \Delta, \tag{6.12}$$

where Δ is a small gap depending on Γ, which can be estimated by WKB method. The third eigenvalue comes from the hybridization of the states of $E_2^{(0)} = \frac{3}{2}\hbar\omega$. In the choice of the parameter (6.9), the energy barrier is the same order of the amount of the quantum fluctuation, $\frac{1}{2}\hbar\omega$, and we find only two levels of (6.12) below the potential barrier.

When the energy barrier becomes large compared to the quantum fluctuation, a different feature appears. To see this, we study a case with smaller quantum fluctuation taking a smaller value of the parameter:

$$\frac{\hbar^2}{2m} = 0.00005. \tag{6.13}$$

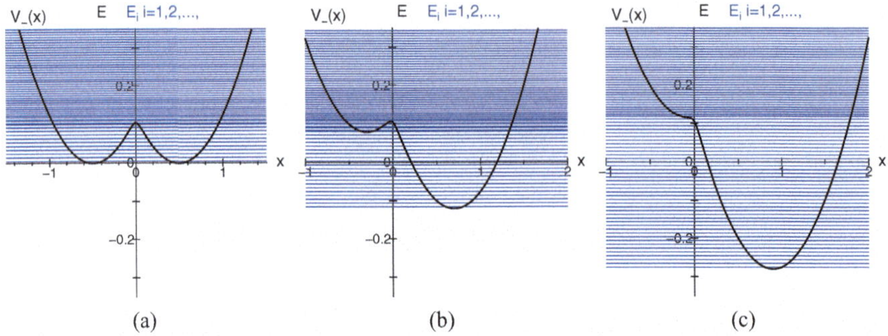

Fig. 6.3 **a** Eigenvalues (horizontal line) and eigenstates (dotted line) for the hybrid potential for $V(x) = (V_1(x) + V_2(x) \pm \sqrt{(V_1(x) - V_2(x))^2 + 4\Gamma^2})/2$ plotted by a bold solid line. Here, $V_1(x)$ and $V_2(x)$ are given in (6.4) with $\Gamma = 0.02$, and $H = 0$, $\hbar^2/2m = 0.00005$. **b** $H = 0.2$, **c** $H = 0.4$

In this case, many states appear below the energy barrier. The levels for $H = 0$ are depicted in Fig. 6.3a. It should be noted that the low energy states below the potential barrier for $H = 0$ are almost doubly degenerate with very small separations which we cannot see, and they overlap in the figure.

Energy structures of systems with nonzero field, $H = 0.2$ and 0.4, are depicted in Fig. 6.2b and c, respectively, for $\hbar^2/2m = 0.005$. In Fig. 6.3b and c, those for $\hbar^2/2m = 0.00005$ are depicted. In the cases of $H \neq 0$, lower energy levels are single (not quasi-degenerate), where the population locates on the right-hand side of the potential. In the case of $H = 0.2$, we find doubly degenerate states in the energy region of metastable part of the potential. The hybridization of states in this region causes the quantum tunneling from metastable well.

Metastability may be discussed in the energy levels depicted in Fig. 6.2a, but in this case the barrier is the same order of the quantum fluctuation $\hbar\omega$ and the relaxation occurs in microscopic time. When we consider the tunneling due to the quantum fluctuation, the barrier is much larger than the quantum fluctuation $\hbar\omega$. In this case, it takes a long time to relax. Thus, we study the second case for the metastability in quantum systems.

6.2.1 *Chracteristics of Metastability in the Eigenstate Spectrum $\{E_i(h)\}$ as a Function of Field*

To catch a characteristic of metastability, we study the dependence of eigenstate spectrum $\{E_i\}$ on the field H. In Fig. 6.4a, the eigenvalues for $\frac{\hbar^2}{2m} = 0.005$ and $k = 1$ are depicted as a function of H.

For a negative H, the left well has lower energy, and the lowest energy state is a localized state in the left well. The energy of this ground state gives the lowest energy

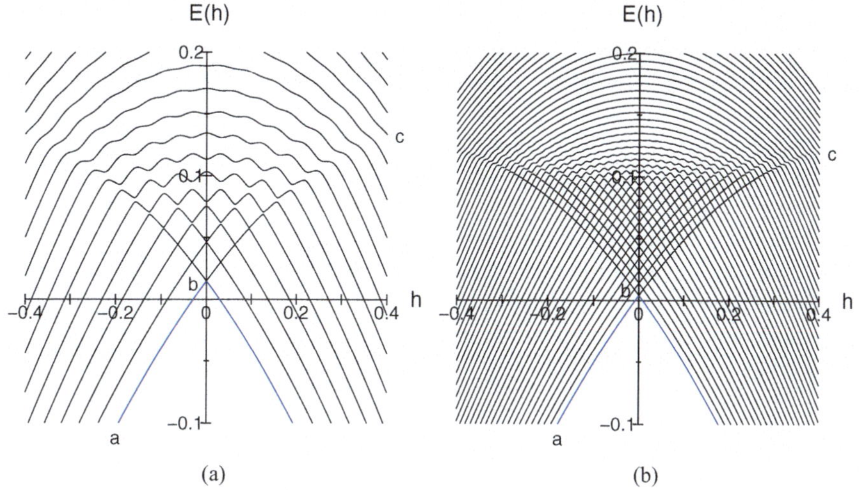

Fig. 6.4 Eigenvalues as functions of H (in the figure $h = H$). $V(x) = (V_1(x) + V_2(x) \pm \sqrt{(V_1(x) - V_2(x))^2 + 4\Gamma^2}/2$ is plotted by a bold solid line. Here, $V_1(x)$ and $V_2(x)$ are given in (6.4) with $\Gamma = 0.02$ and $k = 1$. **a** $\frac{\hbar^2}{2m} = 0.005$. **b** $\frac{\hbar^2}{2m} = 0.0005$

level in the region of $H < 0$. If we sweep the field H from the negative value, the ground state moves along the lowest level in the region $H < 0$ which has a positive slope. Because the Hamiltonian includes the external field $-Hx$, the slope is related to the expectation value in the ith eigenstate $\langle x \rangle_i$ as

$$\frac{dE_i}{dH} = -\langle x \rangle_i. \tag{6.14}$$

Thus, the positive slope, i.e., $\langle x \rangle_i < 0$, indicates that the population localizes in the region of negative x.

When H becomes positive, the true ground state changes to a state localized in the right well, which is shown by the lowest energy level in the right region ($H > 0$). Between these energy levels, an avoided level crossing structure with a very small energy gap is formed at $H = 0$, which is too small to see in the figure.

If the velocity of the field sweep is infinitesimally small, the ground state would adiabatically change from the state localized state in the left well to the state localized state in the right well. Namely, the state moves along the lowest energy levels in Fig. 6.4. However, when the velocity is not very small, the wave function remains in the left well as a metastable state even after H is swept to the positive side. In this motion, the state continues to move along the branch of the energy level which is an extension of the lowest energy level in the region of $H < 0$. Avoided level crossing structures are formed when the extended branch crosses with other levels of negative slope.

The energy gaps at crossing points become large as H increases. We can see the gap at the 4th crossing in Fig. 6.4a. The branch extending to the region of opposite sign of the field indicates the metastable state, and thus we name the extended part *metastable branch*. In Fig. 6.4a, the metastable branch continues to around $H = 0.4$. This point corresponds to the field at which the local minimum of $V(x)$ in the left side disappears as we see in Fig. 6.2.

When $\hbar^2/2m$ becomes small, the energy levels become dense. In the case of $\frac{\hbar^2}{2m} = 0.0005$ the levels are too dense to draw and we depict the levels for $\frac{\hbar^2}{2m} = 0.005$. The structure of metastable branch is seen clearer as depicted in Fig. 6.4b. In the limit $\hbar^2/2m \to 0$, the classical picture of potential trap of particle is reproduced. So far, we studied the case of $k = 1$, but the strength of quantum fluctuation is generally estimated by the value of $\hbar^2 k/2m$.

6.3 Characterstic of Eigenvalue Structure Around the First-Order Phase Transition

From the viewpoint of eigenvalue problem, the metastability is characterized by the picture found above. Namely, the true stable state (the ground state) is an isolated state given by the lowest line in Fig. 6.4. The metastable state is a state on the metastable branch, which is an extension of the stable state after the crossing point. The branch is exposed by many avoided level crossings. At an avoided level crossing point, the state with negative $\langle x \rangle$ and a state with a positive $\langle x \rangle$ are hybridized, and some amount of the population of the metastable branch is scattered to the stable state. This process gives a quantum mechanical relaxation of metastable state in a sweeping field.

We may characterize the stable-metastable transition, i.e., the first-order phase transition, by the following picture depicted in Fig. 6.5. This picture will be applied to a particle conveyance problem in Sect. 6.4 and also to the Stoner–Wohlfarth problem in Sect. 6.10.

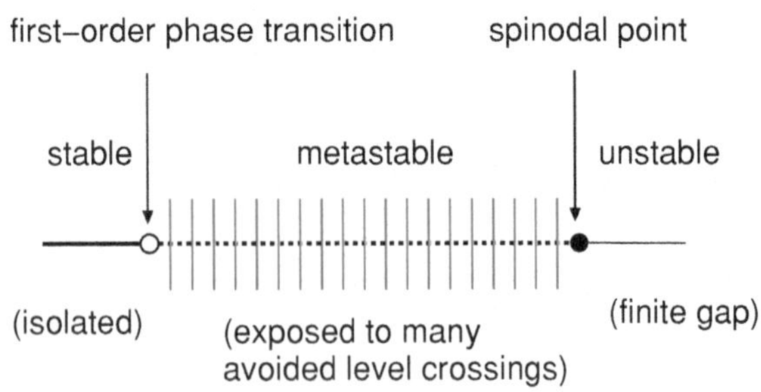

Fig. 6.5 Characterization of the first-order phase transition in eigenvalue problem

The scattering of the population is given by the Landau–Zener formula (see Appendix 8.3) as

$$p \simeq \exp\left(-\frac{(\Delta E)^2}{4\pi v}\right), \tag{6.15}$$

where ΔE is the energy gap at the crossing point and v is the velocity of the field. The effect of appearance of metastable state has been studied with a picture of imaginary eigenvalues for metastability [2–4, 59] as we saw in the previous chapter. The relation between the picture of Fig. 6.5 and the picture of imaginary eigenvalues is an interesting problem, which will be discussed in Sect. 6.4.3.

6.4 Particle Conveyance by a Potential-Well

Now, we consider conveyance of a particle by sweeping the position of a potential-well keeping the particle trapped in it. When the sweeping velocity is small, the particle stays in the potential-well and moves together with the potential-well. However, when the velocity changes, the particle escapes from the potential-well with some probability. For the escape in the process of conveyance of particle, a non-adiabatic change due to the sudden change of Hamiltonian at starting time and quantum tunneling due to metastability of the potential during the conveyance are important issues [65].

The Hamiltonian for a rested potential is given by

$$\mathcal{H}_0 = \frac{1}{2m}p^2 + V(x). \tag{6.16}$$

We consider the potential depicted in Fig. 6.6 for the potential-well $V(x)$:

$$V_{\rm well}(x) = z_p\left[\tanh^2\left(\frac{x}{z_{\rm pw}}\right)\right]. \tag{6.17}$$

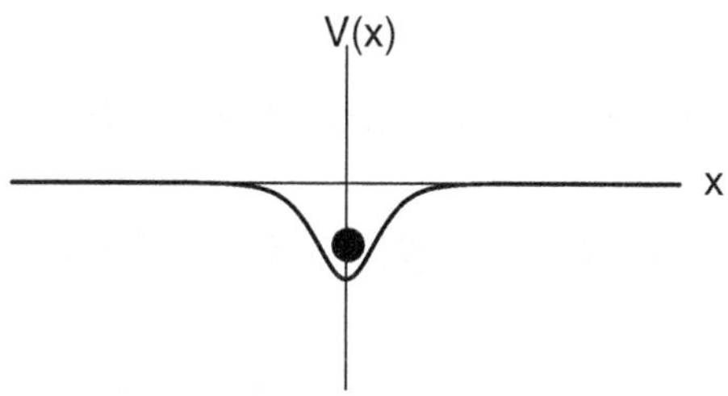

Fig. 6.6 Potential-well for conveyance of particles,

The system has bound states, in which a particle is trapped [60].[2] Here, we use the case of only one bounded state.

When the potential well moves with a velocity c, the potential is time dependent as

$$V(x,t) = z_p \left[\tanh^2 \left(\frac{x - ct}{z_{\mathrm{pw}}} \right) \right], \tag{6.24}$$

and the Hamiltonian becomes time dependent:

$$\mathcal{H}(t) = \frac{1}{2m} p^2 + V(x,t). \tag{6.25}$$

The Schrödinger equation is given by

$$i\hbar \frac{\partial}{\partial t} \Psi(x,t) = \mathcal{H}(x,t)\Psi(x,t). \tag{6.26}$$

In the frame moving with the potential, the wave function is expressed as

$$\Phi(x.t) = e^{ipct/\hbar} \Psi(x,t), \tag{6.27}$$

where

$$p = -i\hbar \frac{\partial}{\partial x}, \quad e^{ipct/\hbar} x e^{-ipct/\hbar} = x + ct. \tag{6.28}$$

---

[2] The Schrödinger equation

$$-\frac{\hbar^2}{2m} \frac{d^2}{dx^2} \psi - \frac{U_0}{\cosh^2(\alpha x)} \psi = E\psi \tag{6.18}$$

is transformed to the form

$$\frac{d}{d\xi} \left[(1 - \xi^2) \frac{d\psi}{d\xi} \right] + \left[s(s+1) - \frac{\varepsilon^2}{1 - \xi^2} \right] \psi = 0 \tag{6.19}$$

by new variables

$$\xi = \tanh \alpha x, \quad \varepsilon = \frac{\sqrt{-2mE}}{\hbar\alpha}, \quad \frac{2mU_0}{\alpha^2\hbar^2} = s(s+1). \tag{6.20}$$

This is the associated Legendre equation, and its solution is given using hypergeometric function as

$$\psi = (1 - \xi^2)^{\varepsilon/2} F \left[\varepsilon - s, \quad \varepsilon + s + 1, \varepsilon + 1, \frac{1 - \xi}{2} \right], \tag{6.21}$$

which has bounded solutions when $\varepsilon - s = -n, n = 0, 1, 2, \ldots$. Then, the eigenvalues of bounded states are

$$E_n = -\frac{\hbar^2\alpha^2}{8m} \left[-(1 + 2n) + \sqrt{1 + \frac{8mU_0}{\alpha^2\hbar^2}} \right]^2, \tag{6.22}$$

where

$$n < s = \frac{1}{2} \left(\sqrt{\frac{8mU_0}{\alpha^2\hbar^2}} - 1 \right). \tag{6.23}$$

.

The Schrödinger equation (6.26) is rewritten as

$$i\hbar\frac{\partial}{\partial t}\Phi(x,t) = (\mathcal{H}_0(x) - cp)\,\Phi(x,t). \tag{6.29}$$

The Hamiltonian in this equation is transformed as

$$\mathcal{H}_1 = \mathcal{H}_0 - cp = \frac{1}{2m}p^2 - cp + V(x) = \frac{1}{2m}(p-mc)^2 - \frac{1}{2}mc^2 + V(x). \tag{6.30}$$

Then, we introduce a unitary transformation

$$f(x,t) = e^{-icmx/\hbar}\Phi(x,t), \quad e^{icmx/\hbar}pe^{-imx/\hbar} = p - cm. \tag{6.31}$$

In this frame, the Schrödinger equation (6.29) is given by

$$i\hbar\frac{\partial}{\partial t}f(x,t) = \left(\mathcal{H}_0(x) - \frac{1}{2}mc^2\right)f(x,t). \tag{6.32}$$

Thus, the Hamiltonian in this frame is

$$\mathcal{H}_2 = \mathcal{H}_0(x) - \frac{1}{2}mc^2. \tag{6.33}$$

The eigenvalues of $\mathcal{H}_2$ are the same as those of $\mathcal{H}_0$ with a constant shift $mc^2/2$. The wave functions are related as

$$f(x,t) = e^{-icmx/\hbar}\Phi(x,t) = e^{-icmx/\hbar}e^{ipct/\hbar}\Psi(x,t). \tag{6.34}$$

6.4.1 Sudden Start by Changing the Velocity from Zero to c

First, we consider the case that the potential-well suddenly begins to move with a velocity c. At the initial time ($t = 0$), the potential stays at the origin and the initial state is the ground state $|0\rangle$ of the system $\mathcal{H}_0$:

$$\mathcal{H}_0|0\rangle = E_0|0\rangle. \tag{6.35}$$

In frame of $f(x,t)$ given in (6.31), at $t = 0$, this state is expressed by the eigenstate of $\mathcal{H}_2$ as

$$f(x,0) = e^{-icmx/\hbar}|0\rangle = \sum_{k=0}^{\infty} c_k|k\rangle, \tag{6.36}$$

where $|k\rangle$ is the kth eigenstate of $\mathcal{H}_0$. Ignoring the constant shift $mc^2/2$, the eigenvalues $\{E_k\}$ of $\mathcal{H}_2$ are the same as those of $\mathcal{H}_0$. The time evolution of the wave function is given by

$$f(x,t) = \sum_{k=0}^{\infty} c_k e^{-iE_k t/\hbar}|k\rangle. \tag{6.37}$$

This indicates that the probability that the particle is trapped in the potential-well after the sudden is given by

$$p_{\text{trapped}} = |c_0|^2. \tag{6.38}$$

If the potential has only one bound state, the wave functions $|k\rangle (k \geq 1)$ are extended states, and the population $1 - |c_0|^2$ escapes from the potential by the non-adiabatic effect due to the shock of the sudden start.

6.4.2 *Smooth Acceleration*

In the previous subsection, we considered a sudden start. Now, we consider a constant acceleration in which the velocity changes from 0 smoothly:

$$v = at, \tag{6.39}$$

where a is a constant acceleration rate.

In this case, the acceleration constant a suddenly changes from 0 to an nonzero value. In this sense, the change is still sudden. But, in the present case the change is smoother than the previous case. The Schrödinger equation is given by

$$i\hbar\frac{\partial}{\partial t}\Psi(x,t) = \mathcal{H}_a(t)\Psi(x,t) = \left[\frac{p^2}{2m} + V\left(x - \frac{1}{2}at^2\right)\right]\Psi(x,t). \tag{6.40}$$

With an unitary transformation

$$U_1 = \exp\left(-i\frac{1}{2\hbar}at^2 p\right), \quad U_1^{-1}\mathcal{H}_a(t)U_1 = \mathcal{H}_0, \tag{6.41}$$

the wave function in the moving frame is

$$\Phi(x,t) = U_1^{-1}\Psi(x,t), \tag{6.42}$$

which follows the Schrödinger equation

$$i\hbar\frac{\partial}{\partial t}\Phi(x,t) = \left(\frac{1}{2m}(p - amt)^2 - \frac{1}{2}m(at)^2 + V(x)\right)\Phi(x,t). \tag{6.43}$$

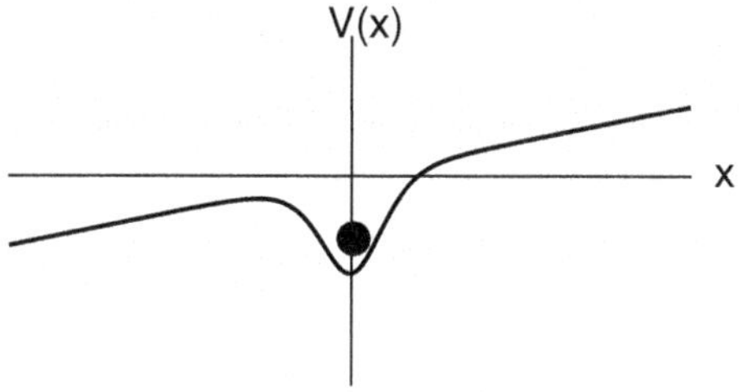

Fig. 6.7 Tilted potential well for conveyance of particles

Here we use another unitary transformation

$$V = e^{-imatx/\hbar}, \tag{6.44}$$

which shifts the momentum

$$V^{-1} p V = p - mat. \tag{6.45}$$

The transformed wave function:

$$\chi(x, t) = V^{-1} \Phi(x, t) \tag{6.46}$$

follows

$$i\hbar \frac{\partial}{\partial t} \chi(x, t) = (\mathcal{H}_0(x) + amx + V(x))\, \chi(x, t). \tag{6.47}$$

Thus, we find that motion of uniformly accelerated particle is equivalent to that in the system with a tilted potential

$$V_a(x) = V(x) + amx, \tag{6.48}$$

which is depicted in Fig. 6.7.

There, the particle will escape to the left by tunneling effect. For a large a, i.e., for a rapid acceleration, the slope is large, and the potential barrier is small. Then, the particle easily escapes from the potential-well. This escape corresponds to particle drop-off from the potential-well during the acceleration.

Let us study the eigenvalues of the system as a function of a. We numerically solve the Schrödinger equation:

$$-\frac{\hbar^2}{2m} \frac{d^2}{dx^2} \psi + (V(x) + amx)\psi = E\psi.$$

By discretization of space and time as $x = j\Delta x$ and solve an eigenvalue problem of an $N \times N$ matrix ($N = 2L/\Delta x$) for the relation which is given by[3]

[3] We may use higher order formulae for numerically efficiency. But, here we used the simplest one.

$$-\frac{\hbar^2}{2m(\Delta x)^2}(\Psi(j+1)-2\Psi(j)+\Psi(j-1))+(V(j\Delta x)+aj\Delta x)\Psi(j)=E\Psi(j) \quad (6.49)$$

Here, we take the parameters as

$$z_p=4, \quad zw=3, \quad \Delta x=0.5, \quad \frac{\hbar^2}{2m(\Delta x)^2}=80. \quad (6.50)$$

When $a=0$, the system has one bounded state $E_1(a=0)\sim -1.9$. In the case of $a>0$, the potential function $V(x)$ is $-\infty$ at $x=-\infty$ and ∞ at $x=\infty$, i.e., $V(\pm\infty)=\pm\infty$. And thus, except near $x=0$, the potential is essentially $V(x)=ax$ for which the wave function is given by the Airy function.

We are interested in the behavior of the trapped particle, and we investigate the eigenvalue structure around zero energy, i.e., around $x=0$. For this purpose, we confine the system within the space $[-200, 200]$ as depicted in Fig. 6.8a. The potential functions around $x=0$ are magnified in Fig. 6.8b.

In Fig. 6.9, the lowest 30 eigenstates $\{E_i\}$, $i=0,1,2\cdots 29$ are plotted as a function of a. In the figure, at $a=0$, we find a bound state at $E_0\sim -1.9$ and extended excited states with positive energies E_i, $(i\geq 1)$. The eigenvalues change with a. For $a>0$, the potential well becomes metastable, where the flat line representing the metastable bound state crosses with the extended states. Here, we find a sequence of avoided level crossings. This structure has the same structure with the metastable branch as that in Fig. 6.5. Because we confine the system in a finite range $[-L, L]$, $L=200$, for small a, the system has a bound state in Fig. 6.9, but, when L becomes large, the density of lines becomes dense and the bound state disappears.

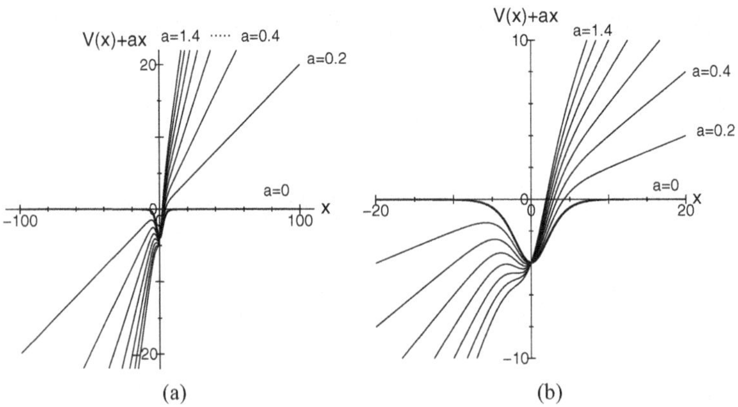

Fig. 6.8 **a** Potential functions $V(x)-ax$ used for the numerical calculation, which are confined in $[-200, 200]$. In the figure the part $[-100, 100]$ is drawn. **b** Magnified view around $x=0$

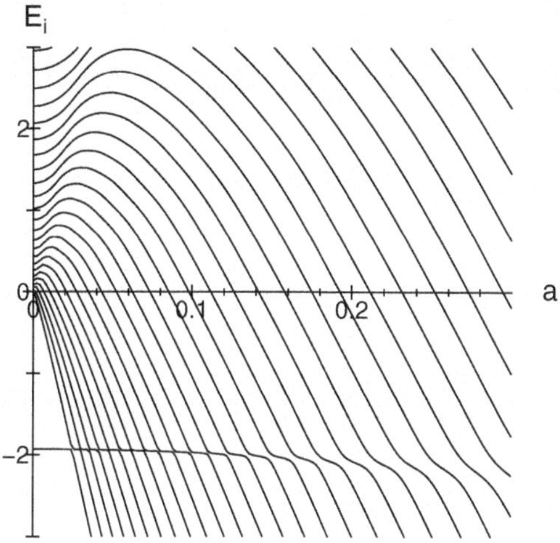

Fig. 6.9 The lowest 30 eigenstates $\{E_i\}, i = 1, 2 \cdots 30$

When we increase L, $V(-L)$ at the left edge becomes lower than the trap depth. Then, the escape of a particle from the well to the left region occurs even for a small a. Thus, the bound state in energy structure at small a is a consequence of a finite L. In the present model, true stable trapped state exists only at $a = 0$. Namely, if L becomes infinity, the flat line is the metastable branch for all values of a except $a = 0$. This means, when we accelerate the potential-well, the population inevitably escapes. But the rate is very small.

When a increases further and the trap potential has no more local minimum, beyond which the system does not have metastable branch.

In the present analysis, we used a system confined in a finite region, and thus all the eigenstates are bounded although we called the states with positive eigenenergy 'extended states' which correspond to a continuum spectrum in the case $L = \infty$. Thus, we do not have conceptual difficulty for the normalization of the wave function. But this fact restricts the time period in which we can properly discuss dynamics of the particle. That is, precisely speaking, the dynamics of trapped population in the present model is applicable only in a short initial time, i.e., before the change of the wave function reaches to the boundary. If the escaping particle reaches the boundary of the model, the present treatment is no more adequate. However, as long as we concern 'metastable state', this early time analysis can capture important characteristics of relaxation.

6.4.2.1 Relaxation Dynamics

Next, we study the real-time dynamics of a wave function in the potential with $a > 0$. The initial state of the wave function is the ground state of the trapping potential at $a = 0$.

$$\Psi(t) = e^{-i\mathcal{H}_a t/\hbar}\Psi(0), \quad \mathcal{H}_{a=0}\Psi(0) = E_0^{a=0}\Psi(0). \tag{6.51}$$

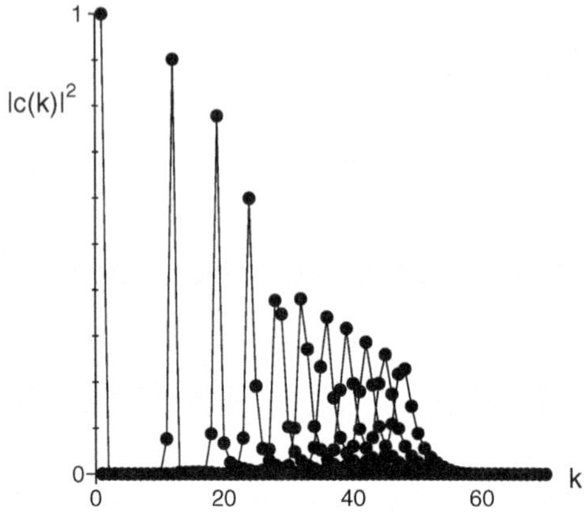

Fig. 6.10 Distributions of $|c(k)|^2$ in systems with $a = 0, 0.1, 0.2, \ldots, 1.0$. $|c(k)|^2$, $(k = 0, 1, \ldots)$ of the same a are connected by a thin line

The population of the state $\Psi(0)$ relaxes in time, because it is no more eigenstate of the system $\mathcal{H}_a$. We study the relaxation for several values of a.

The initial state is expanded by the eigenstates of $\mathcal{H}_a$, $\{|\phi_i^a\rangle\}$:

$$\Psi(0) = \sum_k c_k |\phi_k^a\rangle, \tag{6.52}$$

and $\Psi(t)$ evolves as

$$\Psi(t) = \sum_k c_k e^{-E_k^a t/\hbar} |\phi_k^a\rangle, \tag{6.53}$$

where $E_k^a$is the kth eigenvalue of $\mathcal{H}_a$.

Distributions of $|c(k)|^2$ in the systems with $a = 0, 0.1, 0.2, \ldots, 1.0$ are depicted in Fig. 6.10. For $a = 0$, $|c(k)|^2 = \delta_{k,0}$ as it should be. For $a > 0$, the distribution becomes wider as a increases. The peaks locate around $E_k^a \sim E_0^{a=0}$ which gives the metastable branch. Although there are 400 bases, $|c(k)|^2$ above 60th eigenstates are almost zero and not shown in the figure.

In Fig. 6.11, we depict the overlap of the wave function and the initial one:

$$X = |\langle\Psi(0)|\Psi(t)\rangle|^2 = \left|\sum_k c_k^* \langle\phi_k^a| \sum_m c_m e^{-iE_m^a t/\hbar} |\phi_m^a\rangle\right|^2$$

$$= \sum_k \sum_m |c_k|^2 |c_m|^2 \cos((E_m^a - E_k^a)t/\hbar). \tag{6.54}$$

Here we find decays of the overlap $X(t)$. This relaxation is a purely quantum mechanical phenomenon. In the present case, the initial state is the ground state which is a bound state in a trapped potential-well. At $t = 0$, we suddenly accelerate it with a constant acceleration constant a. Thus, a change of frame occurs, and the initial state

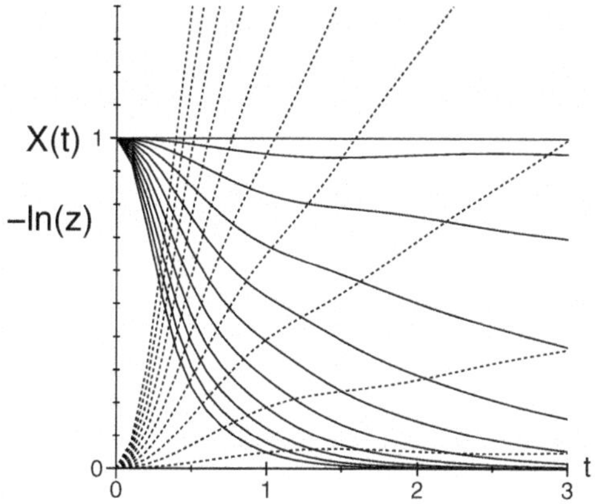

Fig. 6.11 Decay of the overlap $X(t) = |\langle\Psi(0)|\Psi(t)|^2$, for $a = 0, 0.1, \ldots, 1.0$. from top to the bottom. Dotted line shows $-\ln(X(t))$

is expanded by the eigenstates as (6.52). Because all the states are extended state for $a > 0$ and the populations spread in the space, and the overlap decays.

At late time in the figure, the relaxations seem to be an exponential decay. To see this fact, we also plot $-\ln X(t)$, which shows two regions. From the slop at late time, we estimate the relaxation time τ:

$$X(t) \sim e^{-t/\tau}. \tag{6.55}$$

The sharper slope indicates a faster relaxation time. We may have the following intuitive pictures. The populations of eigenstates whose energies are out of trapping potential decay fast, which gives the fast initial slope. The populations of eigenstate localized in the trap decay by quantum tunneling which gives a slow exponential decay. This relaxation rate can be estimated by a model adopting an absorbing boundary condition. In the model, we have imaginary eigenvalues, and the imaginary part gives the exponential decay of overlap.

Here, the dynamics in the pure quantum dynamics is given by (6.54). The apparent exponential decay may come from the summation over many frequencies $\{2\pi(E_m^a - E_k^a)/\hbar\}$ as an decoherent phenomena. For example, if the density of states $D(\omega)$ of the energy gap $E_m^a - E_k^a \equiv \hbar\omega$ has a Lorentzian form:

$$D(\omega) = \frac{1}{a^2 + \omega^2}, \tag{6.56}$$

then, the dynamics is given by

$$\int_0^\infty \cos(\omega t) D(\omega) d\omega = \frac{\pi}{2a} e^{-at}. \tag{6.57}$$

Indeed, it is found that the density $D(\omega)$ of $\{E_m^a - E_k^a\}$ has the form (6.56) [77]. Relation between the resonant state and the population distribution over the eigenstates of the Hamiltonian is an interesting problem [30, 77].

If we use a deeper potential-well, there are several bound states and several resonance states with different relaxation rates appear. In such cases, the relaxation of the overlap can be given by sum of exponential decays [78]. In the present study, we used the potential-well with only one bound state, thus the fast relaxation is due to the dynamics above the energy barrier.

6.4.3 Scattering Problem

This type of problem has been studied as problem of the potential scattering, where the idea of 'resonance state' or the so-called non-Hermitian Hamiltonian method has been developed [30, 59, 66–74]. When only outgoing wave function is assumed, the conservation of probability is violated and Hermite property is violated. This idea has been used to explain natural radiation decay of exited state.

The mechanism of relaxation of resonance state is explained in detail in the textbook [30]. Here we briefly explain the mechanism by making use of WKB method (see Appendix 8.6).

6.4.4 Relaxation from Metastable Potential

Scattering problem has been studied for the potential depicted in Fig. 6.12. Setting

$$p(x) = \sqrt{2m(E - V(x))}, \quad \rho(x) = \sqrt{2m(V(x) - E)}, \tag{6.58}$$

wave functions in regions are

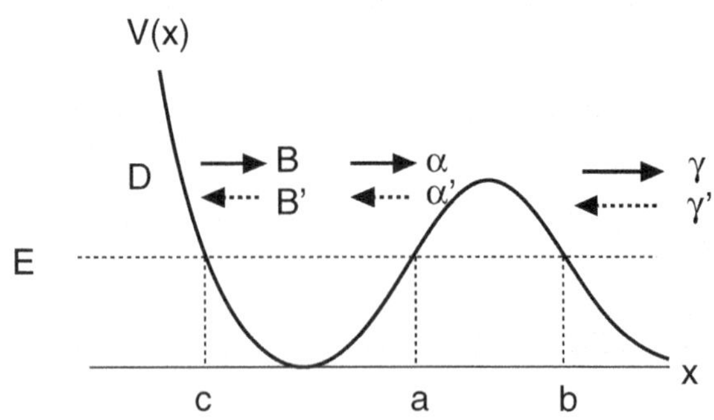

Fig. 6.12 Typical shape of potential $V(x)$ for the scattering problem

$$\psi(x > b) = \frac{\gamma}{\sqrt{p(x)}} \exp\left[\frac{i}{\hbar}\int_b^x p(x')dx'\right] + \frac{\gamma'}{\sqrt{p(x)}} \exp\left[-\frac{i}{\hbar}\int_b^x p(x')dx'\right], \tag{6.59}$$

$$\begin{aligned}\phi(c < x < a) &= \frac{\alpha}{\sqrt{p}} \exp\left[\frac{i}{\hbar}\int_a^x p(x')dx'\right] + \frac{\alpha'}{\sqrt{p}} \exp\left[-\frac{i}{\hbar}\int_a^x p(x')dx'\right] \\ &= \frac{B}{\sqrt{p}} \exp\left[\frac{i}{\hbar}\int_c^x p(x')dx'\right] + \frac{B'}{\sqrt{p}} \exp\left[-\frac{i}{\hbar}\int_c^x p(x')dx'\right]\end{aligned} \tag{6.60}$$

and for $x < c$, the wave function must exponentially damp, and we put

$$\psi(x < c) = D\frac{1}{\sqrt{\rho(x)}} \exp\left[-\frac{1}{\hbar}\int_x^a \rho(x')dx'\right]. \tag{6.61}$$

Using the connection formula explained in Appendix 8.6, the wave functions in $x > b$ are given by

$$\begin{pmatrix}\gamma \\ \gamma'\end{pmatrix} = \begin{pmatrix}\left(e^S + \frac{e^{-S}}{4}\right) e^{-i\frac{\pi}{4}} e^{iX} - i\left(e^S - \frac{e^{-S}}{4}\right) e^{i\frac{\pi}{4}} e^{-iX} \\ i\left(e^S - \frac{e^{-S}}{4}\right) e^{-i\frac{\pi}{4}} e^{iX} + \left(e^S + \frac{e^{-S}}{4}\right) e^{i\frac{\pi}{4}} e^{-iX}\end{pmatrix}, \tag{6.62}$$

where

$$S = \frac{1}{\hbar}\int_a^b \rho(x')dx', \quad X \equiv \frac{1}{\hbar}\int_c^a p(x')dx'. \tag{6.63}$$

Now, we introduce new variables r and θ as

$$e^{-i\frac{\pi}{4}} r e^{i\theta} = \left(e^S + \frac{e^{-S}}{4}\right) e^{-i\frac{\pi}{4}} e^{iX} - i\left(e^S - \frac{e^{-S}}{4}\right) e^{i\frac{\pi}{4}} e^{-iX} \tag{6.64}$$

where

$$r^2 = 2e^{2S} + \frac{e^{-2S}}{8} + \left(2e^{2S} - \frac{e^{-2S}}{8}\right)\cos(2X). \tag{6.65}$$

and

$$\theta = \mathrm{Arg}\left(\frac{\left(e^S + \frac{e^{-S}}{4}\right) e^{iX} + \left(e^S - \frac{e^{-S}}{4}\right) e^{-iX}}{r}\right). \tag{6.66}$$

Thus,

$$\begin{pmatrix} \gamma \\ \gamma' \end{pmatrix} = Dr \begin{pmatrix} e^{i(\theta - \frac{\pi}{4})} \\ e^{-i(\theta - \frac{\pi}{4})} \end{pmatrix}. \tag{6.67}$$

When

$$\cos(2X) = -1, \tag{6.68}$$

the relation for the condition of the bound state (see Appendix 8.6.3)

$$\frac{1}{\hbar} \int_c^a p(x')dx' = 2n\pi + \pi \tag{6.69}$$

holds. In this case,

$$r^2 = \frac{e^{-2S}}{4}, \tag{6.70}$$

i.e., γ^2 is very small. Therefore, the population in the metastable valley is large and the state represents the metastable state. On the other hand, when $\cos(2X) \neq -1$, $\gamma \propto e^{2S}$, and thus the population on the metastable valley is very small. That is, the wave function at $x > b$ is mostly reflected at the point b.

Near the energy E_n which satisfies $\cos(2X) = -1$, we set

$$E = E_n + \Delta E, \tag{6.71}$$

and we expand X as

$$X(E) = \left(\frac{1}{2} + n\right) + \frac{\pi}{\hbar\omega}\Delta E. \tag{6.72}$$

Here $\hbar\omega$ is a constant which corresponds to the energy gap $E_{n+1} - E_n$. Substituting this, we have

$$r(E)^2 \simeq \frac{e^{-2S}}{4} + \left(2e^{2S} - \frac{e^{-2S}}{8}\right)\frac{1}{2}\left(\frac{2\pi}{\hbar\omega}\Delta E\right)^2 \simeq \frac{e^{-2S}}{4} + e^{2S}\left(\frac{2\pi}{\hbar\omega}\Delta E\right)^2. \tag{6.73}$$

Thus,

$$r^{-2} = \frac{\hbar\omega}{2\pi} \frac{\Gamma_n}{(E - E_n)^2 + \Gamma_n^2}, \quad \Gamma_n = \frac{\hbar\omega}{2\pi} e^{-2S}, \tag{6.74}$$

which has a Lorentzian form. This energy distribution indicates an exponential decay of the initial wave function in time in pure quantum dynamics.

6.4.4.1 Resonance State

Using outgoing boundary condition [66], it is known that the energy becomes complex. Let us assume $\gamma' = 0$, i.e., only outgoing wave exists, the following relation holds:

$$2e^S \cos(X) - i\frac{e^{-S}}{2}\sin(X) = 0. \tag{6.75}$$

This equation does not have real solution of the energy, and then we put

$$E = E_n - i\frac{\Gamma_n}{2}. \tag{6.76}$$

Substituting (6.72), to (6.75), we have

$$2e^S \frac{\pi}{\hbar\omega}\Delta E \sim -i\frac{e^{-S}}{2}. \tag{6.77}$$

Thus,

$$E = E_n - i\frac{\hbar\omega}{4\pi}e^{-2S}, \tag{6.78}$$

which gives (6.74)

$$\Gamma_n = \frac{\hbar\omega}{2\pi}e^{-2S}.$$

This state with the imaginary energy is called 'resonance state'.

6.4.4.2 Connection Using Parabolic Cylinder Equation

We have obtained the wave function over the barrier region. In this case, we have two turning points. Here, we obtained a wave function over the potential barrier using the connection formula for a single turning point twice which is good for high barrier. However, more sophisticated connection formula over the potential barrier is also known, which is obtained by making use of the asymptotic form of the parabolic cylinder equation:

$$\left(\frac{d^2}{dx^2} + \frac{1}{4}x^2 - a\right)W(x) = 0, \tag{6.79}$$

and it works well for low barrier [30]. There, we still find Lorentian distribution, which gives the reason why we find exponential decays even for large values of a in Fig. 6.11.

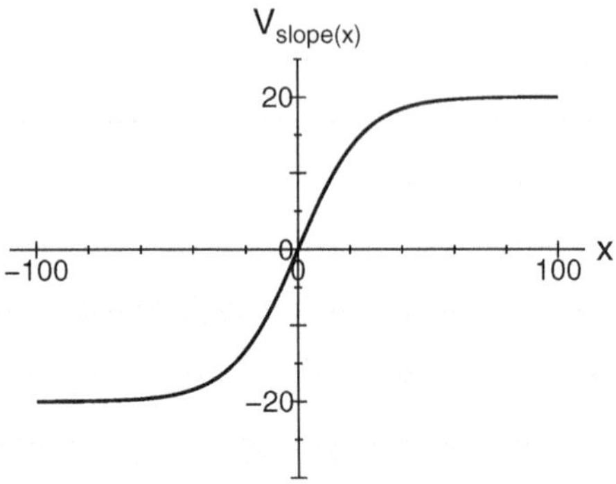

Fig. 6.13 A potential $V_{\mathrm{slope}}(x)$ with a slope between low energy region (left) and high energy region (right)

6.4.5 Carry Up the Particle

Next, we carry up a particle from a low potential region to a high potential region in the potential $V_{\mathrm{slope}}(x)$ depicted in Fig. 6.13.

The potential energy for the particle is given by $V_{\mathrm{slope}}(x) + V(-x_0 + ct)$ which has the form depicted in Fig. 6.14. Thus, at the slope, the potential function is similar to that we studied in the previous subsection.

In Fig. 6.15, the energy levels as a function of the position of the potential-well $x = -x0 + ct$ are depicted. There, we find a sequence of avoided level crossings. The line going up with x represents the metastable trapped state, which again has the same structure as that of metastable branch (Fig. 6.5). In this case, the particle being carried up is exposed to the sequence of avoided level crossings, which is a characteristic of metastability.

When the position of potential well $x = -x_0 + ct$ is swept, the trapped population moves along the metastable branch. When it goes through an avoided level crossing structure of energy levels, some population moves along an energy level adiabatically, which causes a drop-off from the potential. The probability of the adiabatic transition $p_{\mathrm{adiabatic}}$ is estimated by Landau–Zener formula [84–86] explained in Appendix 8.3, which is given by

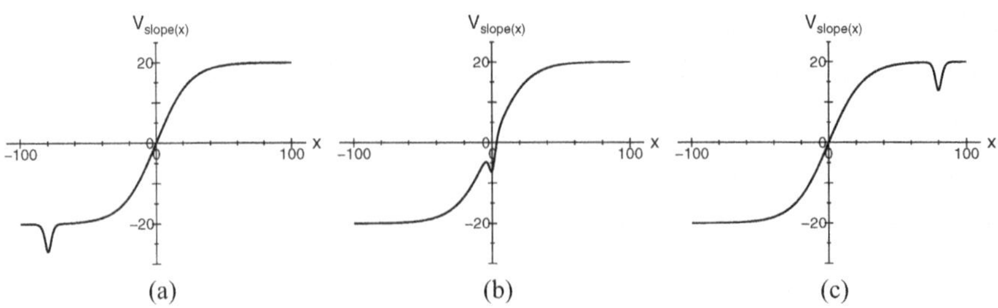

Fig. 6.14 **a** Potential of $V_{\mathrm{slope}}(x) + V(x + x_0)$, **b** $V_{\mathrm{slope}}(x) + V(x)$, **c** $V_{\mathrm{slope}}(x) + V(x - x_0)$

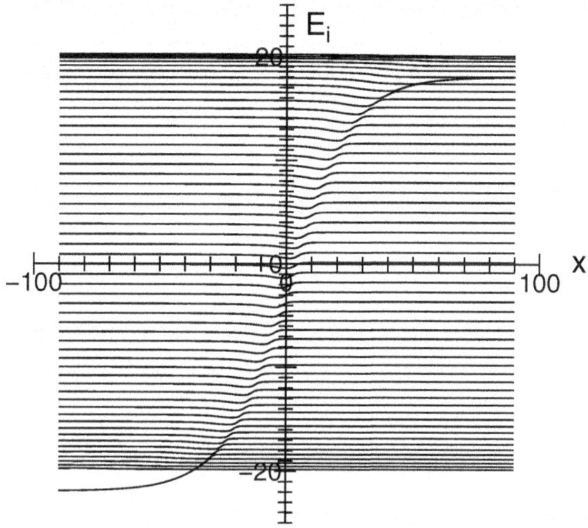

Fig. 6.15 Energy levels when the potential well is located at $x = -x_0 + ct$. Here, $x_0 = 90$

$$p_{\text{adiabatic}} = 1 - \exp\left(-\frac{\pi(\Delta E)^2}{4v}\right), \tag{6.80}$$

where ΔE is the energy gap at the avoided level crossing. The motion along the metastable branch is the non-adiabatic transition and the population is given by $p_{\text{non-adiabatic}} = 1 - p_{\text{adiabatic}}$. This formula naturally explains the dependence of the drop rate on the velocity.

If we sweep the potential very slowly, the particle escapes from the potential well by quantum tunneling. To sweep fast, we need a large velocity, which causes drop-off from the potential due to sudden change of the velocity as we saw in (6.38). Thus, there is a competition between both effects, and there is an optional way to sweep for the conveyance [65]. This type of competition has been pointed out as problem of an annealing schedule for quantum annealing [79].

6.4.5.1 Adiabatic and Non-adiabatic

Here, we note the meanings of the word "adiabatic" in the classical system and in the quantum system. This word originally means that there is no heat exchange. In quasistatic process, this word means no entropy change. Then, it means that a state stays continuously in the same energy level during the process. For example, when a parameter changes slowly, the state which is initially in the ground state stays in the ground state of the current Hamiltonian at the current time, which is the adiabatic theorem. And this type of change is called adiabatic.

The word 'adiabatically' means differently in classical and quantum systems when we discuss the metastable state. In a classical system, if the parameter changes slowly, the state ‘adiabatically’ stays in metastable state until the spinodal point. On the other hand, in a quantum system, the energy level of metastable state crosses with various energy levels where they form an avoided level crossing. In this case, adiabatic change means not staying in the metastable state but staying in the same energy levels, which brings the system out of the metastable state. Thus, the process staying in the metastable state is non-adiabatic in the quantum system, which is opposite to that in the classical systems. This is just a problem of wording, but occasionally it causes a confusion.

6.5 Quantum Tunneling in Magnetic Systems

So far, we studied metastability of a particle in a potential with an energy barrier. In this section, we study metastability in magnetic systems. Magnets are materials with uniaxial anisotropy of magnetization which show coercivity.

6.5.1 *Metastability in Magnetic Systems*

To describe magnetic system microscopically, we express the system with spin operators, which satisfy the commutation relation:

$$[S^x, S^y] = i\hbar S^z, \quad [S^y, S^z] = i\hbar S^x, \quad [S^z, S^x] = i\hbar S^y. \tag{6.81}$$

Usually, we express the state with the basis which is an eigenstate of S_z

$$S^z|m\rangle = m|m\rangle, \quad m = -S, -S+1, \ldots, S, \tag{6.82}$$

where S represents the length of spin and takes values $S = 1/2, 1, 3/2, \ldots$ and the relation

$$\left((S^x)^2 + (S^x)^2 + (S^x)^2\right)|m\rangle = S(S+1)|m\rangle \tag{6.83}$$

holds. The so-called localized spin system is expressed by a generalized Heisenberg model:

$$\mathcal{H}_{\text{localized spin system}} = -\sum_{i=1}^{N}\sum_{j=1}^{N}\sum_{\alpha=x,y,z}\sum_{\beta=x,y,z} J_{ij}^{\alpha,\beta} S_i^\alpha S_j^\beta$$

$$-\mu_{\rm B}\sum_{i=1}^{N}\left(g_x H_x S_i^x + g_y H_x S_i^y + g_x H_z S_i^z\right) - \sum_{i=1}^{N}\sum_{\alpha=x,y,z}\sum_{\beta=x,y,z} D^{\alpha,\beta} S_i^\alpha S_i^\beta, \tag{6.84}$$

where S_i^α is the α component of the spin at ith site ($\alpha = x, y, z, i = 1, \ldots, N$), $J_{ij}^{\alpha,\beta}$ is the exchange coupling, g_α is a constant called 'g-factor', and $D^{\alpha,\beta}$ is the single-ion anisotropy. Higher order anisotropies also exist due to the crystal-electric-field (CEF), i.e., the Stevens operators [80], which are not shown here. The exchange coupling term can be expressed in the following form taking the principal axis of the interaction:

$$\sum_{\alpha=x,y,z} \sum_{\beta=x,y,z} J_{ij}^{\alpha,\beta} S_i^\alpha S_j^\beta = J_x S_i^x S_j^x + J_y S_i^y S_j^y + J_z S_i^z S_j^z$$

$$+ d_x(S_i^y S_j^z - S_i^z S_j^x) + d_y(S_i^z S_j^x - S_i^x S_j^z) + d_z(S_i^x S_j^y - S_i^y S_j^x). \tag{6.85}$$

The former is the symmetric term, and the latter is anti-symmetric term with respect to exchange of i and j. The latter is called 'Dzyaloshinskii-Moriya interaction'. Usually we call (J_x, J_y, J_z) 'Heisenberg exchange coupling' and (d_x, d_y, d_z) 'Dzyaloshinskii-Moriya interaction constant'.

In the following, we consider a simple case

$$\mathcal{H} = -\sum_{i=1}^{N}\sum_{j=1} \left(J_x S_i^x S_j^x + J_y S_i^y S_j^y + J_z S_i^z S_j^z \right) - g\mu_B \sum_{i=1}^{N} H_z S_i^z - D \sum_{i=1}^{N} (S_i^z)^2. \tag{6.86}$$

For uniaxial magnets, we have the following two cases:

$$\text{single} - \text{ion anisotropy model} \quad J_x = J_y = J_z, \quad D > 0 \tag{6.87}$$

and

$$\text{anisotropic coupling model} \quad J_z > J_x \geq J_y, \quad D = 0. \tag{6.88}$$

The extreme case of the latter model with $J_x = J_y = 0$ is the Ising model.

In the thermodynamic limit, these systems have a phase transition in high dimensions $d \geq 2$. In the ordered phase, the magnetization exhibits a first-order phase transition at $H = 0$ as explained in Chap. 2.

In a finite system, due to the anisotropy, the magnetization still shows a microscopic metastable phenomenon with a finite relaxation time. In the next section, we study such phenomena.

6.6 Relaxation of Magnetism in Small Systems

In a small magnetic system with the uniaxial anisotropy, the energy barrier ΔE between the two stable states is finite and the relaxation time of the metastable state is finite. But the system still shows a hysteresis phenomenon in a finite-time operation. The stability (we may call it 'coercivity') is affected by thermal and also quantum fluctuation. The thermal fluctuation causes the Arrhenius type relaxation time, i.e., $1/\tau \sim e^{-\beta\Delta E}$ and the relaxation time becomes very long at low temperatures. The thermal fluctuation causes the super-paramagnetism for systems of small ferromagnetic particles when the temperature is about the energy barrier. At temperatures below which the super-paramagnetism is suppressed, spins are frozen. This temperature is called 'blocking temperature'. However, even below the blocking temperature, quantum fluctuation may play an important role.

Observations of temperature independent relaxation time at low temperatures are often attributed to the quantum tunneling. But they are often owing to distribution of system size [63]. Thus, measurement of single nanoparticle is important which have been done by Grenoble group [81]. Some symptoms of quantum effect were discussed, but no clear evidence was observed for nano-size particles which consist of thousands of atoms.

However, clear evidence of quantum tunneling was observed in the so-called single molecular magnets (SMM) [82] which is a crystal of almost magnetically isolated molecules. The magnetization curve of the crystal can be regarded as the expectation value of the magnetization of each molecule. Quantum tunneling is also observed in the magnetization process of rare-earth ions contained as impurities in matrices [83]. The ions are really isolated there.

6.7 Single Molecular Magnets (SMM)

As examples of SMM, crystals consisting of molecules of Mn_{12} and Fe_8 have been discovered [82]. The magnetic Hamiltonians of these molecules are expressed by a spin system of $S = 10$ with easy-axis type anisotropy:

$$\mathcal{H} = -D(S_z)^2 - HS_z + Q, \tag{6.89}$$

where the eigenvalues of S_z are

$$M_z = -S, -S+1, \ldots, S. \tag{6.90}$$

Q is a term which does not commute with S_z and causes quantum mixture of states with different M_z. The energy level structure as a function H for a small amplitude of Q has a similar structure to that of $Q = 0$, in which the energy is a function of $m = M_z$ as

$$\mathcal{H}|m\rangle = E_m|m\rangle. \quad E_m = -Dm^2 - Hm \tag{6.91}$$

which is depicted in Fig. 6.16. The crossing point of E_m and $E_{m'}$ is given by

$$-Dm^2 - Hm = -D(m')^2 - Hm' \rightarrow (m-m')(D(m+m')+H) = 0. \tag{6.92}$$

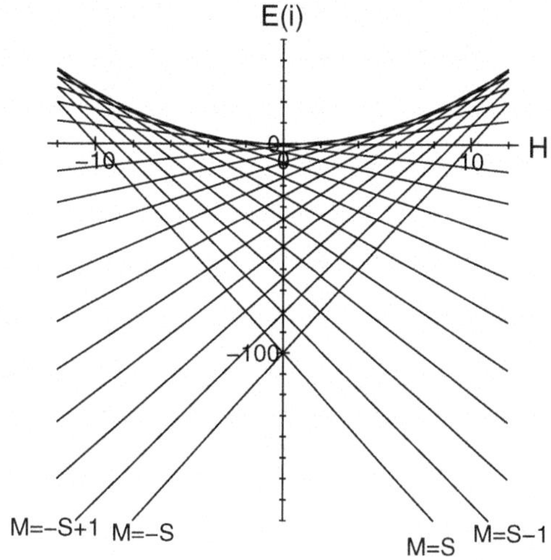

Fig. 6.16 *H* dependence of energy levels for $S = 10, D = 1, Q = 0$

Thus, they cross at the field

$$H_c = -D(m + m'). \tag{6.93}$$

If $Q = 0$, the state of $M_z = m$ is an eigenstate and it does not change when H changes. At the crossing point of two states of $M_z = m$ and m', they simply cross without hybridization.

On the other hand, when $Q \neq 0$, the states $\{|m\rangle\}$ are no more eigenstate, and the eigenstates $\{|\phi_i\rangle\}$ are given by linear combinations of $\{|m\rangle\}$:

$$|\phi_i\rangle = \sum_{m=-S}^{S} c_{i,m}|m\rangle. \tag{6.94}$$

However, when Q is small the eigenstates are close to $\{|m\rangle\}$, i.e.,

$$c_{i,m} \simeq \delta_{i,m}, \tag{6.95}$$

except at crossing point. Thus, they may be still labeled by m.

In Fig. 6.17a, we depict the eigenstates as functions of H for

$$Q = H_x S_x \tag{6.96}$$

with $H_x = 5$. There, we find that the low energy part is similar to Fig. 6.16. However, the energy levels form an avoided level crossing structure at crossing points with an energy gap ΔE.

Here, we used a large value of $H_x (= 5)$ to make ΔE visible. Even so, we cannot see the gap at crossing points of levels $m = S$ and $m' = -10, -9, \ldots, -3$. We barely see the gap at the crossing of ($S = 10$, $m' = -2$) which is marked by a circle.

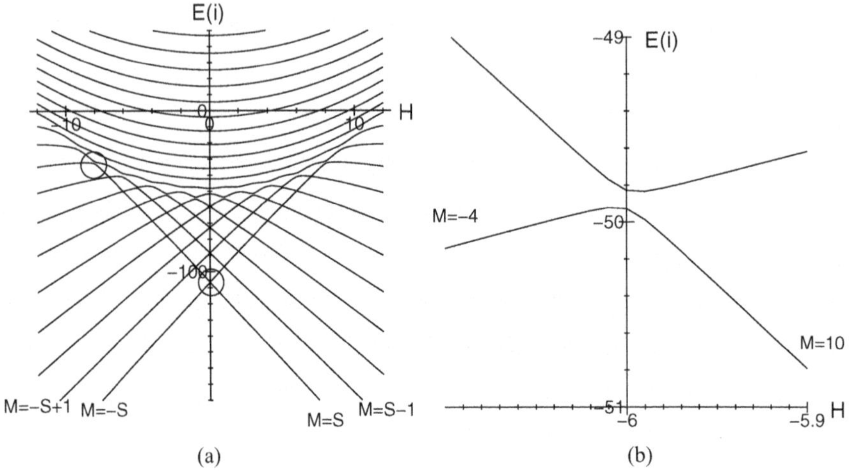

Fig. 6.17 **a** H dependence of energy levels for $S = 10$, $D = 1$, $H_x = 5$ **b** Magnified view for the crossing of $(S = 10, m' = -4)$

A magnified view for the crossing of $(S = 10, m' = -4)$ is depicted in Fig. 6.17b. In this scale, we can see the gap.

Near the crossing point of the levels of m and m', the system can be regarded as a two-state model and its Hamiltonian is given by

$$\mathcal{H}_{m,m'} = \begin{pmatrix} E_m & \delta \\ \delta^* & E_{m'} \end{pmatrix} = E_c + \begin{pmatrix} \Delta m(H - H_c)/2 & \delta \\ \delta^* & -\Delta m(H - H_c)/2 \end{pmatrix} \quad (6.97)$$

where

$$E_m = -Dm^2 - Hm, \quad H \simeq H_c, \quad \delta = \langle m|Q|m'\rangle \quad (6.98)$$

and $E_c = (E_m + E_{m'})/2$ and $\Delta m = (m - m')$. Eigenvalues of this Hamiltonian are

$$\lambda_{\pm} = E_c \pm \sqrt{\left(\frac{\Delta m(H - H_c)}{2}\right)^2 + |\delta|^2}. \quad (6.99)$$

The energy gap is $2|\delta|$.

When the field H is swept through the crossing point, the state is scattered into the two levels because of the quantum mixing term Q. In the initial time ($H = \infty$), let the initial state be at the level m. Then, the field is swept with velocity $-v$. The probability that the state is in the level m' in the final state is given the Landau-Zener formula [84–88]. (see Appendix 8.3).

$$p_{m \to m'} = 1 - \exp\left(-\frac{\pi(\Delta E)^2}{2|(m - m')v|}\right). \quad (6.100)$$

The process of $m \to m'$ is called adiabatic transition because the state adiabatically stays in the same level, by which magnetization changes from $m > 0$ to $m' < 0$. On the other hand, the process of $m \to m$ is called non-adiabatic transition because the state jumps from the original level to the other level, by which magnetization does not change. If $\Delta E \simeq 0$, i.e., $Q \simeq 0$, $p_{m\to m'} \simeq 0$. If the sweeping velocity v is very large, $p_{m\to m'}$ is also close to 0. This means that the most population moves on the level with the same slope m non-adiabatically. In the opposite case with a large ΔE and small v, the state moves along the same level adiabatically and changes the magnetization $m \to m'$. In this case, $p_{m\to m'} \simeq 1$. This formula was used to analyze the transition probability at crossing points in experiments [89, 90].

When we sweep the magnetic field slowly from a large positive value, the state starting from the ground state ($m = S$) moves along the level until H reaches 0. There, it crosses with the state of $m = -S$ and then $m = -S + 1 \ldots$, and so on. At the crossing point with the level of m', the population $P(S)$ of the state of $m = S$ is reduced by $P(S)p_{S\to m'}$ and the change of magnetization is

$$\langle \Delta M \rangle = -P(S)p_{S\to m'}(S - m'), \quad m' = -S, -S + 1, \ldots, \tag{6.101}$$

although the field is still below the value where the energy barrier disappears (spinodal point). This change of magnetization due to the quantum mixing at a crossing point is called 'resonance tunneling'. The magnetization shows a jump at the crossing as

$$\langle M \rangle \to \langle M \rangle - P(S)p_{S\to m'}(S - m'). \tag{6.102}$$

When H is apart from crossing points, the state moves along the level of $m = S$ and the magnetization does not change. Thus, in the sweeping process, the magnetization largely changes only near each crossing point, which causes a step-like magnetization curve. In Fig. 6.18, examples of the changes of magnetizations under sweeping velocities of the fields, $v = 1, 0.1, 0.01, 0.001$ and 0.0001, are depicted. When the velocity becomes slow, the change of magnetizations at small fields becomes visible because $p_{S\to m'}$ becomes large for small v.

6.7.1 Tunneling Under Dissipation

In the pure quantum mechanical dynamics, the population scattered to the state m' stays at the same state after the crossing point. However, in real situations of experiments, the scattered population to m' relaxes to the ground state $M_z = -S$ immediately owing to the thermal effect as depicted in Fig. 6.19. Namely, the population of tunnels to the left side relaxes to the ground state. Therefore, the amount of change of magnetization is not (6.101), but is given by [91]

$$\langle \Delta M \rangle = -p_{S\to m'}(S - (-S)), \quad (m' = S, -S + 1, \ldots). \tag{6.103}$$

This relation explains the experimental data of the change of magnetization [90].

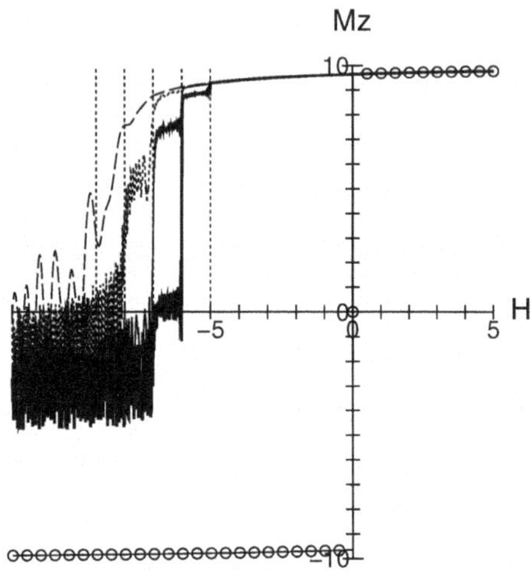

Fig. 6.18 Changes of magnetizations $M_z(t)$ under sweeping fields $H(t) = 5 - vt$ with the velocities $v = 1$(dashed line), 0.1(dotted line), 0.01(thin solid line) and, 0.001(bold solid line). The magnetization in the ground state is plotted by circles which changes from S to $-S$ at $H = 0$. This sudden change would occur in the limit $v \to 0$

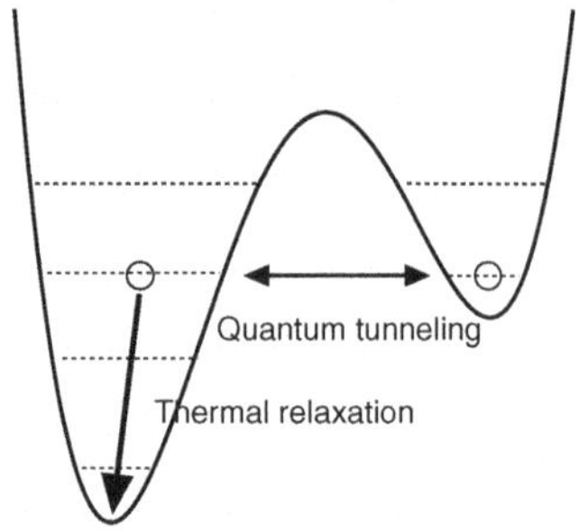

Fig. 6.19 Relaxation to the ground state after tunneling,

6.7.2 Dynamics in Dissipative Environments

To study the effects of dissipation, we need to take into account contacts with the thermal bath, which is explained in Appendix 8.4. Dynamics in dissipative environments is studied by a quantum master equation for the density matrix ρ:

$$i\hbar\frac{\partial}{\partial t}\rho = [\mathcal{H}, \rho] + \gamma D[\rho], \tag{6.104}$$

where $D[\rho]$ causes relaxation process which can be given by Lindblad form (Appendix 8.4.1), Redfield form (Appendix 8.4.2), etc. In Fig. 6.20, we depict magnetization curve, in the same way in Fig. 6.18, in dissipative environments under a sweeping field.

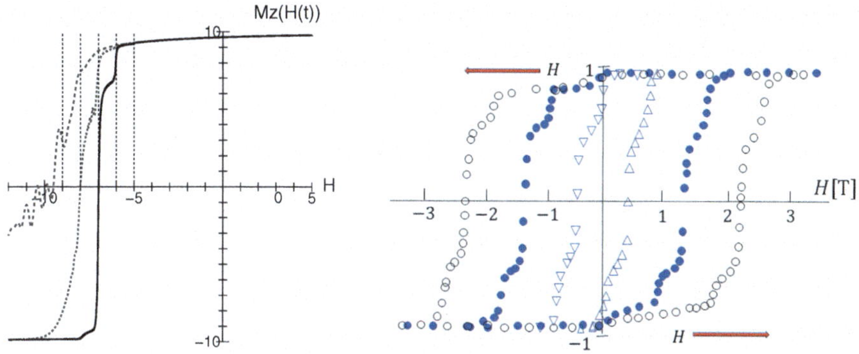

(a) (b)

Fig. 6.20 **a** Example of magnetization curve under sweeping field $v = 0.1, 1$, and 10 with a dissipation ($S = 10$, $D = 1, H_x = 5$). **b** Schematic picture of magnetization curve of Mn_{12} at low temperatures, $T = 1.5$K(open circles), $T = 1.9$K(closed circles), $T = 2.4$K(downward triangle for the temperature decreasing process, upward triangle for the temperature increasing process). Courtesy Bernard Barbara

6.7.2.1 Thermally Assisted Resonance Tunneling

It should be noted that in experiments at finite temperatures the populations at excited states $0 < m < S$ are nonzero due to the thermal fluctuation. The populations of the excited states also exhibit the Landau–Zener transition to the state m'' at the crossing fields $H_c = D(m + m'')$ which are the same value for the crossing point of $m = S$. The gap at the quasi-crossing of m and m'' is generally larger than that of S and m', and thus $p_{m \to m''} > p_{S \to m'}$. Thus, even when $p_{S \to m'}$ is too small to be observed, the magnetization change due to the crossing of (m, m'') may have a large amount and be observed. This tunneling between excited states is called 'thermally assisted resonance tunneling'. This type of step-like magnetization, indeed, has been observed in experiments of SMM [82].

6.8 Magnetic Foehn Effect

Dynamics of magnetization under a sweeping field with both quantum and thermal fluctuation shows interesting behavior. Here, we consider a simple two-level system:

$$\mathcal{H} = -H(t)\sigma^z - \Gamma\sigma^x, \quad H(t) = vt. \tag{6.105}$$

Here we sweep the field from negative to positive side, starting from the ground state with the initial negative field. Quantum mechanically, when the field goes through the avoided level crossing point $H(t) = 0$, some amount of population is scattered by the Landau–Zener mechanism (Appendix 8.3). The magnetization $m \simeq -1$ changes at the crossing point. The magnetization after the crossing is given by

$$m_{\rm LZ}(v) \sim 1 - 2p_{\rm non-adiabatic}, \quad p_{\rm non-adiabatic} = 1 - \exp\left(-\frac{\pi\Gamma^2}{v}\right), \tag{6.106}$$

and $m_{\rm LZ}(v)$ becomes large when v decreases.

In the motion in dissipative environment, the dynamics is given by the quantum master equation explained in Appendix 8.4. At finite temperatures, if the system well contacts with the thermal bath, the magnetization for a given temperature T and a field H is given by the isotherm magnetization:

$$m_T(H,T) = \frac{H}{\sqrt{H(t)^2+\Gamma^2}} \tanh\left(\beta\sqrt{H(t)^2+\Gamma^2}\right). \tag{6.107}$$

On the other hand, the adiabatic magnetization which is the magnetization of the ground state at the given H is

$$m_{\rm ad}(H) = \frac{H}{\sqrt{H^2+\Gamma^2}}, \tag{6.108}$$

which corresponds to the pure quantum case with $v = 0$. If we sweep the magnetic field very slowly, the magnetization changes as $m_{\rm ad}(H,T)$ at zero temperature, while it changes as $m_T(H,T)$ at a finite temperature T.

When the velocity v is large and the effect of contact with the thermal bath is neglected during the crossing, the magnetization follows the LZ formula, and the magnetization just after the crossing is given by $m_{\rm LZ}$. After the crossing, the magnetization gradually relaxes to the equilibrium value by the effect of contact with the thermal bath. When the velocity v becomes small, quantum dynamics of $m(t)$ approaches $m_{\rm ad}(H)$. This dependence is depicted in Fig. 6.21a, where the magnetization just after the crossing increases when the velocity decreases.

On the other hand, for a very slow sweep, the effect of contact with the thermal bath during the crossing becomes relevant, and some population of the ground state is excited to bring the system to the thermal equilibrium value $m_T(H,T)$, i.e.,

$$m(t) = m_{\rm ad}(H) \rightarrow m_T(H,T). \tag{6.109}$$

In this slow sweep case, the magnetization shows a plateau. The magnetization at the plateau decreases when v decreases in contrast to the case of Landau–Zener mechanism for the fast sweep. This dependence is depicted in Fig. 6.21b. The derivative $dm(t)/dt$ shows a peculiar double-peak structure. The latter phenomenon is understood as a process in which heat injects to the system during an adiabatic motion, and we called them 'magnetic Foehn effect' [92].

In many realistic experimental systems, the sweeping velocity is slow and the latter case is dominated. This effect was found in the experiment of Fe_2 [93]. This

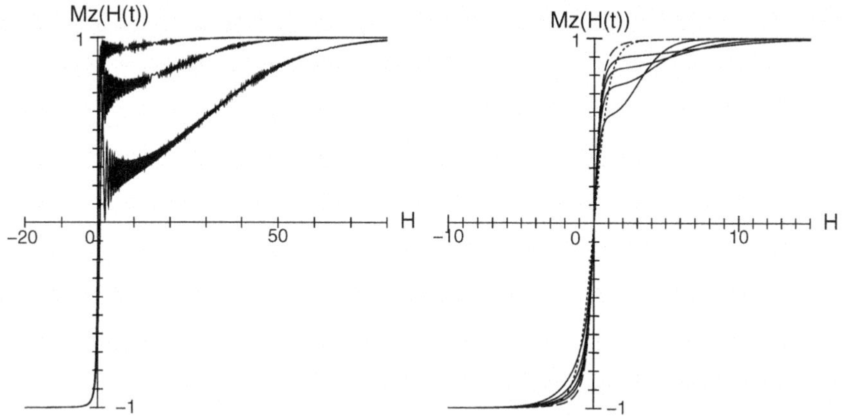

Fig. 6.21 **a** Dynamics of the magnetization for $\Gamma = 0.5$, $\gamma = 0.001$. **a** Under fast field sweeps: $v = 0.4$, 0.2, and 0.1 from the bottom to top. **b** Under slow field sweeps: $v = 0.01$, 0.005, and 0.002. from the top to bottom. The dotted line denotes the isothermal magnetization (6.107), and the dashed line the adiabatic magnetization (6.108)

type of plateau is also found in magnetization curves obtained by the pulse field on Fe_{10} [94]. In a slow sweep, the magnetization process is given by $m_T(H, T)$. But if the heat capacity of the reservoir is small, it cannot supply heat enough. In this case, the magnetization has a plateau. This phenomenon in V_{15} was explained as the phonon bottleneck effect [95].

The noise effect on the LZS transitions have been studied extensively [96–105]. Please find the original works.

6.9 Effect of Dissipation on the Relaxation of Metastable State

6.9.1 Free-Boson Bath Model

Quantum dynamics in the double-well potential has been treated as a two-level system consisting of ground states of both wells with a small gap [75]. The gap corresponds to the tunneling rate between the wells. The effect of contacts with the thermal bath has been studied by using a set of hormonic oscillators [106].

Caldeira and Leggett [76] studied a system of a particle coupled with free-boson bath. The Lagrangian using imaginary time, i.e., Euclidean Lagrangian, is given by

$$L_{\mathrm{E}} = \frac{1}{2}M\dot{q}^2 + V(q) + \frac{1}{2}\sum_{\alpha} m_\alpha \dot{x}_\alpha^2 + \frac{1}{2}\sum_{\alpha} m_\alpha \omega_\alpha^2 x_\alpha^2 + q\sum_{\alpha} c_\alpha x_\alpha. \quad (6.110)$$

Because of the bilinear form of $\{x_\alpha\}$, we can integrate out the bath degree of freedom $\{x_\alpha\}$, and the action is given as

$$S_{\text{eff}} = \int_0^T dt \left[\frac{1}{2}M\dot{q}^2 + V(q)\right] - \int_{-\infty}^{\infty}\int_0^T dt dt' X(t-t')q(t)q(t'), \tag{6.111}$$

where

$$X(t-t') = \sum_\alpha \frac{c_\alpha^2}{4m_\alpha\omega_\alpha} e^{-\omega_\alpha|t-t'|}. \tag{6.112}$$

They introduced a spectrum density $J(\omega)$ as

$$X(t-t') = \frac{1}{2\pi}\int_0^\infty d\omega J(\omega) e^{-\omega|t-t'|}. \tag{6.113}$$

To make a classical motion obey the equation of motion with a friction in the form

$$M\ddot{q} = -\frac{dV(q)}{dq} - \eta\dot{q}, \tag{6.114}$$

$J(\omega)$ was taken to be the form for small $\omega \le \omega_0$

$$J(\omega) = \eta\omega. \tag{6.115}$$

In this case, the effective action is given as

$$S_{\text{eff}} = \int_0^T dt \left[\frac{1}{2}M\dot{q}^2 + V(q)\right] + \frac{\eta}{4\pi}\int_{-\infty}^{\infty}\int_0^T dt dt' \left(\frac{q(t)-q(t')}{(t-t')^2}\right)^2. \tag{6.116}$$

Because the second term is positive, it is concluded that the free-boson effect causing a friction in classical dynamics in the form (6.114) causes reduction of tunneling rate. Bray and Moore discussed an over-damped relaxation due to the friction [107].

6.9.2 *Dynamics of the Magnetization in Uniaxial Anisoropy*

Here, we study dynamics of the magnetization in uniaxial anisotropy which was studied in Sect. 6.7

$$\mathcal{H} = -D(S_z)^2 - H_x S_x, \quad S > \frac{1}{2} \tag{6.117}$$

by a simple Redfield-type master equation. In this section, we study the case of $H_z = 0$. For a small H_x, the states with $M_z = \pm S$ hybridize, and the lowest two eigenstates are given by

$$|1\rangle \simeq \frac{|S\rangle + |-S\rangle}{\sqrt{2}}, \quad |2\rangle \simeq \frac{|S\rangle - |-S\rangle}{\sqrt{2}} \tag{6.118}$$

with the eigenvalues E_1 and E_2, respectively. The energy gap is $\Delta E = E_2 - E_1$.

To study the relaxation of magnetization, we investigate the dynamics of the magnetization starting from the state which is almost $|S\rangle$:

$$|t = 0\rangle = |+\rangle = \frac{|1\rangle + |2\rangle}{\sqrt{2}}. \tag{6.119}$$

Because

$$|t\rangle = \frac{|1\rangle + e^{-i\Delta Et/\hbar}|2\rangle}{\sqrt{2}}, \tag{6.120}$$

the magnetization oscillates between S and $-S$ with a period

$$T = \frac{2\pi\hbar}{\Delta E}. \tag{6.121}$$

For a large S and a small H_x the energy gap is very small [6, 108] as

$$\Delta E \propto \left(\frac{H_x}{2D}\right)^{2S}, \tag{6.122}$$

and thus, the change of magnetization is very slow. This change is regarded as the relaxation due to quantum tunneling over the barrier.

When we study relaxation of the magnetization at low temperatures, we consider the lowest two energy levels which we considered above. Next, we study the effects of dissipation on this process.

6.9.3 *Effects of Dissipation on the Hybridized Lowest Two States*

The effect of dissipation has been studied for the spin-boson model with $S = 1/2$ [109–111]. Here, we study the effect in systems with larger spins.

We use the Redfield type of quantum master equation (8.166) in which the dissipation makes the state relax to the equilibrium state of the Hamiltonian (see Sect. 8.4.2):

$$\frac{\partial \rho(t)}{\partial t} = \left[\mathcal{H}_{\text{tot}}, \rho(t)\right] - \kappa\left([X, R\rho(t)] + [X, R\rho(t)]^{\dagger}\right). \tag{6.123}$$

We may consider the coupling to the bath with $X = S_x$, S_y and S_z. In the case of NMR, the magnetic field is applied, and we take the direction of the field along the z axis. There, the coupling with the transverse components of spin, i.e., $X = S_x, S_y$, causes a spin flip and thus causes energy dissipation, which is called 'transverse relaxation'. On the other hand, the coupling with the longitudinal component of spin, i.e., $X = S_z$, causes the phase diffusion, which is called 'longitudinal relaxation'. Both relaxations play important role.

In the case of large spin $S > 1/2$, the lowest two state may be expressed by an effective $S = 1/2$ system. However, the effect of noise is largely modified due to the energy barrier between the two states $|S\rangle$ and $|-S\rangle$ as has been pointed out in the literatures [75].

To study this situation, we investigate a model of (6.117) with a large S directly. In this model, the couplings to the bath with $X = S_x$, S_y and S_z should be treated equally. Setting the initial state to be a linear combination of the ground state and the first excited state (6.119) in which the magnetization is almost S. Without the dissipation effect, the magnetization shows a sinusoidal motion of the period (6.121). If we study the system of $S = 10$, the energy gap is very small, and the simulation is numerically difficult. Thus, here we use systems with $S = 2$ and $S = 3$ where the energy gap is small but modest.

We depict the motion of the magnetization for a system of $S = 2$ with $D = 0.1$ and $H_x = 0.1$ with $\kappa = 0.0025$ at $T = 0.1$ in Fig. 6.22a. In this case, the eigenvalues are -0.8744563, -0.8605551, -0.4000000, -0.1394449, and 0.2744563, and the energy gap is $\Delta E = 0.01390114$. In the figure, we find damping-oscillations due to the dissipations in both cases of $X = S_y$ and S_z. However, the effect of $X = S_y$ is weaker than that of $X = S_z$.

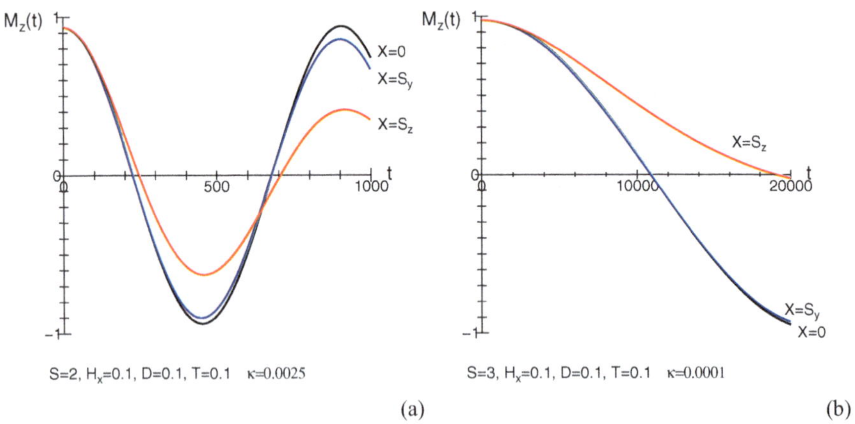

Fig. 6.22 **a** Damping oscillations starting from $(|1\rangle + |2\rangle)/\sqrt{2}$ of a single spin system of $S = 2$ with the damping constant $\kappa = 0.0025$ at temperature $T = 0.1$: $X = S_z$, $X = 0$ (no dissipation), $X = S_y$. **b** $S = 3$ with the damping constant $\kappa = 0.0001$ at temperature $T = 0.1$: $X = S_z$, $X = 0$ (no dissipation) $X = S_y$

This difference between the cases of $X = S_y$ and $X = S_z$ becomes significant in the system of $S = 3$ (Fig. 6.22b) where the gap is much smaller. In the simulation, we use the parameters: $D = 0.1$ and $H_x = 0.1$ with $\kappa = 0.0001$ at $T = 0.1$. Here, the eigenvalues are $-2.790453, -2.790020, -1.406820, -1.322147, -0.6549035, -0.08783307$, and 0.6521761, and the energy gap is $\Delta E = 0.0004320878$. In this case, the effect of $X = S_y$ is almost suppressed, while the effect of $X = S_z$ remains.

The reason for the difference between $X = S_y$ and $X = S_z$ is understood by a consideration on the matrix elements X. For the case of $X = S_y$, the matrix element

$$\langle 1|S_y|2\rangle \simeq \frac{\langle S|S_y|S\rangle - \langle S|S_y|-S\rangle + \langle -S|S_y|S\rangle - \langle -S|S_y|-S\rangle}{2} = 0, \quad (6.124)$$

because S_y can change the magnetization only by 1. The matrix element is of order of the energy gap. This fact is very different from the case of $S = 1/2$ in which the matrix element is $i/2$.

On the other hand, for the case of $X = S_z$, the matrix element is given by

$$\langle 1|S_z|2\rangle \simeq \frac{\langle S|S_z|S\rangle - \langle S|S_z|-S\rangle + \langle -S|S_z|S\rangle - \langle -S|S_z|-S\rangle}{2} = S. \quad (6.125)$$

Thus, only the case of $X = S_z$ causes a relevant dissipation effect for systems of large S. From this observation, when we model the relaxation of magnetization by the lowest two states, only the longitudinal type of noise is relevant.

When the constant a of the coupling $X = aS_z$ changes, the magnetization dynamics changes from a damping oscillation to an over-damped type relaxation [109–112]. The change has been found to be a Kosterlitz–Thouless type phase transition.

6.10 Quantum Stoner–Wohlfarth Model

The Hamiltonian (6.89) with $Q = -H_z S_x$:

$$\mathcal{H} = -D(S_z)^2 - HS_z - H_z S_x, \quad M_z = -S, -S+1, \ldots, S, \quad (6.126)$$

is has the same form as Stoner-Wohlfarth model in the classical spin model $\boldsymbol{m}$

$$\mathcal{H}_{\rm SW} = -\hat{D}(m_z)^2 - Hm_z - H_z m_x, \quad m_x^2 + m_y^2 + m_z^2 = 1, \quad (6.127)$$

which is the simplest model for the metastable state.

Thus, how the singularity at the Stoner–Wohlfarth point appears in the limit $S \to \infty$ in quantum system is an interesting problem [113]. To study this problem, we consider a model with a scaled spin

$$s_x = \frac{S_x}{S}, \quad s_y = \frac{S_y}{S}, \quad s_z = \frac{S_z}{S}, \tag{6.128}$$

for which, the commutation relations are

$$[s_x, s_y] = \frac{i\hbar}{S} s_z, \quad [s_y, s_z] = \frac{i\hbar}{S} s_x, \quad [s_z, s_x] = \frac{i\hbar}{S} s_y. \tag{6.129}$$

The Hamiltonian is written as

$$\mathcal{H} = S\left(-DS(s_z)^2 - Hs_z - H_z s_x\right) \equiv S\left(-D_{\mathrm{SW}}(s_z)^2 - Hs_z - H_z s_x\right), \tag{6.130}$$

where

$$D_{\mathrm{SW}} = DS. \tag{6.131}$$

Comparing (6.130) to (6.127),

$$\mathcal{H} = S\mathcal{H}_{\mathrm{SW}} \tag{6.132}$$

with $\hat{D} = D_{\mathrm{SW}}$. Thus, in the classical system, the time scale in quantum system should be scaled by S, and the sweeping rate c is also scaled:

$$t_{\mathrm{SW}} = \frac{t}{S}, \quad v = Sc \tag{6.133}$$

and the anisotropy constant is also scaled by S (6.131).

The Heisenberg equations of spins are given by

$$\begin{aligned} \frac{ds_x}{dt_{\mathrm{SW}}} &= \left\{D_{\mathrm{SW}}(s_y s_z + s_z s_y) + H_z s_y\right\} \\ \frac{ds_y}{dt_{\mathrm{SW}}} &= \{D_{\mathrm{SW}}(s_z s_x + s_x s_z) + H_x s_z - H_z s_x\} \\ \frac{ds_z}{dt_{\mathrm{SW}}} &= -H_x s_y. \end{aligned} \tag{6.134}$$

Thus, the corresponding time evolution in the classical system is the torque equation:

$$\frac{d\boldsymbol{m}}{dt_{\mathrm{SW}}} = -\boldsymbol{m} \times \boldsymbol{H}_{\mathrm{eff}}, \quad \boldsymbol{H}_{\mathrm{eff}} = \begin{pmatrix} H_x \\ 0 \\ 2D_{\mathrm{SW}} m_z + H_z \end{pmatrix}, \tag{6.135}$$

which agrees with (6.134) ignoring the commutation relations between spin operators, i.e., $s_z s_x + s_x s_z \to 2s_z s_x$. When S becomes large, quantum mechanical motion is expected to approach the classical one in the scaled time scale.

Now, we study the eigenvalues as a function of H for systems with large S. In Fig. 6.17a, we depicted the eigenvalue levels of $S = 10$. In Fig. 6.23a, b and c, we depict the eigenvalue levels of $S = 20$, 40 and 80, respectively. Here, we again find

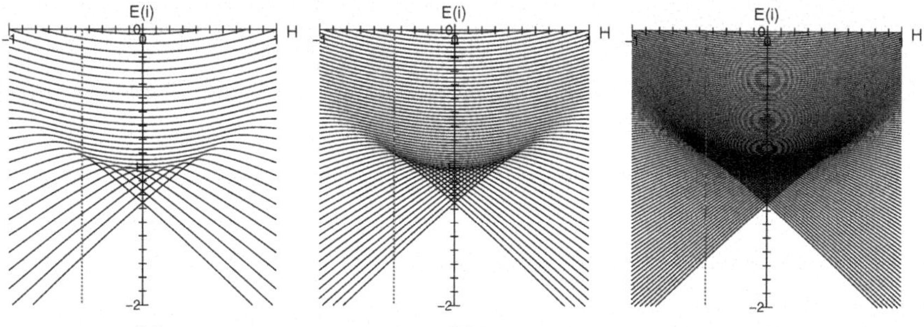

Fig. 6.23 Eigenvalue levels as a function of H of systems with large spins: **a** $S = 20$, **b** $S = 40$ and **c** $S = 80$. $D_{\rm SW} = 1$, $Hx = 1$

a metastable branch structure. The energy levels become dense with S, but a similar structure is maintained. The dotted line denotes the Stoner-Wohlfarth point which is the end of metastability in the classical model.

Now, we study dynamics of populations of each level under a sweeping field with a velocity v ($H(t) = H_0 - vt$). The initial state is the ground state of the system at a large positive $H = H_0$.

$$i\hbar\frac{\partial\Psi(t)}{\partial t} = \mathcal{H}(H(t))\Psi(t), \quad \mathcal{H}(H_0)\Psi(0) = E_0\Psi(0). \tag{6.136}$$

The state is scattered at every avoided level crossing point in $H \leq 0$. An example of change of populations $\{p_i\}$ of states are depicted in Fig. 6.24.

$$p_i = |\langle i(H)|\Psi(t)\rangle|^2, \quad \mathcal{H}(H)|i(H)\rangle = E_i|i(H)\rangle. \tag{6.137}$$

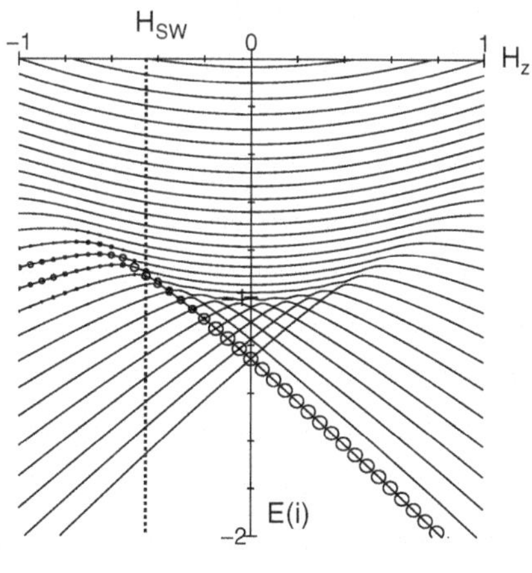

Fig. 6.24 Population distribution in the field-sweeping process. The radius of each circle denotes the population of the state. $S = 20$, $D_{\rm SW} = 1$, $H_x = 1$ $v = 0.004$

At each avoided level crossing point the scattering occurs, and the surviving probability through the kth crossing point is given by the LZ formula

$$p_k = \exp\left(-\frac{\pi(\Delta E_k^{(s)})^2}{2\hbar\Delta m_k^{(s)} c}\right) = \exp\left(-\frac{\pi(\Delta E_k^{(s)})^2}{2\hbar\Delta m_k^{(s)}(v/S)}\right), \tag{6.138}$$

where $\Delta E_k^{(s)}$ is the energy gap at the kth crossing point which locates at $H = H_k^{(s)}$ and v is the velocity of the field. As long as S is finite, the relaxation occurs little by little at the crossing points and no singularity appears as we see in Sect. 6.7. When S is large, p_k is expected to be close to 1 for $|H| < H_{\rm SW}|$ and has a singularity at $H_{\rm SW}$.

Now, we study how the singularity appears in the limit $S \to \infty$. The difference of the magnetizations at the kth crossing levels, at which levels of S and of $-S + (k-1)$ cross, is

$$\Delta m_k^{(s)} \simeq \frac{2S-k}{S}. \tag{6.139}$$

As S becomes large, the number of the eigenvalues increases as $2S+1$. Thus, if we increase S to $2S$, a new eigenvalue appears at every interval of the eigenvalues of the system of S as shown in Fig. 6.25. In Fig. 6.25a, level crossings of states of m and m' and $(m'+1)$ as depicted. When S becomes double ($S \to 2S$), the number of states is also doubled, and an extra state appears in every interval.

The surviving of probability in this period of H in the system of S should be the same that in the system of $2S$ for large S and thus

$$p_k(S) \simeq p_{2k}(2S)p_{2k}(2S). \tag{6.140}$$

To satisfy this condition, the gap should be scaled

$$\Delta E_{m,m'}(S) = 2\Delta E_{2m,2m'}(2S). \tag{6.141}$$

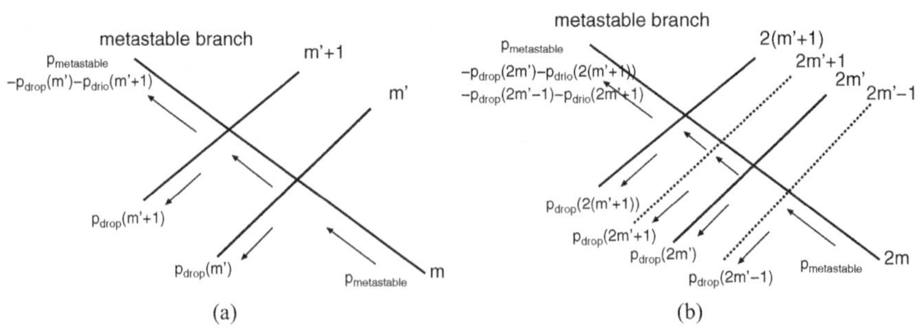

Fig. 6.25 Schematical pictures of crossings of energy levels in **a** a system with spin S and **b** spin $2S$

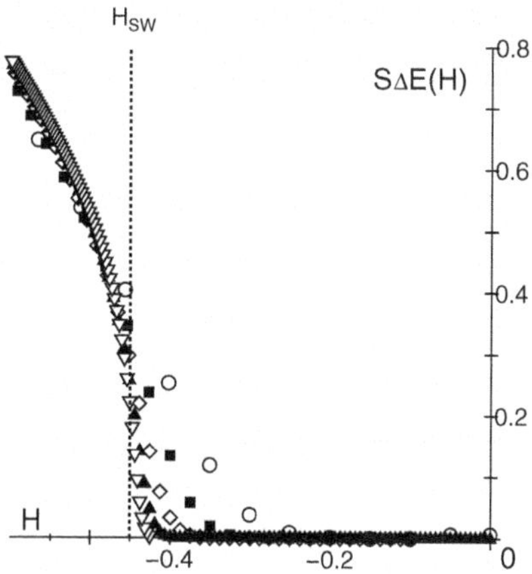

Fig. 6.26 The energy gaps multiplied by S as a function of H for various values of S. $S = 20, 40, 80, 160$ and 320. $D_{\rm SW} = 1$, $H_x = 1$. The dotted line denotes the Stoner-Wohlfarth point $H_{\rm SW}$

With similar analysis, we find that the energy gap must have a scaling property in the form [113]:

$$\Delta_{\rm scaled} E = S \Delta E(S). \tag{6.142}$$

When this property is satisfied, the survival probability in an interval in the metastable branch in the limit $S \to \infty$ exists naturally.

In Fig. 6.26, the scaled gaps $\Delta_{\rm scaled} E$ for various values of S are plotted as a function of the field H of crossing points. For $H > H_{\rm SW}$, we find that the scaled gap decreases with S and vanishes in the limit $S \to \infty$, which corresponds to the metastability in the classical model. For $H < H_{\rm SW}$ the scaled gap converges to a curve. The limiting curve suggests a singular dependence of energy gap for the Stoner-Wohlfarth criticality. In this region, the magnetization in the classical model begins to rotate around the direction of field as seen in Fig. 2.8.

For the dependence of the curve on S, we may plot the data in a scaling plot in analogy to the classical spinodal point [23], regarding S as the system size:

$$(S\Delta E)^2 \propto S^{-1/3} g\left((H - H_{\rm SW})S^{2/3}\right). \tag{6.143}$$

There, we find a good collapse of data in a single line [113] (not shown) which suggests the scaling works.

6.10.1 Dynamics of Magnetization

Let us see dynamics of the magnetization under a swept field. In Fig. 6.27, we show an example of $(m_x(t), m_y(t), m_z(t))$ of the system (2.16) with $\Gamma = 1$ and $H_x = 1$ and $v = 0.08$. There, we find peculiar beatings [113, 114].

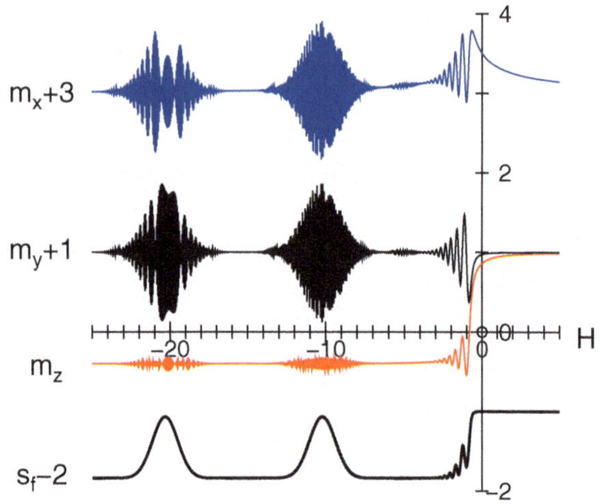

Fig. 6.27 Magnetization curve $(m_x(t), m_y(t), m_z(t))$ under sweeping field in pure quantum mechanical dynamics with $\Gamma = 1$ and $H_x = 1$ and $v = 0.08$. $m_x(t)$ and $m_y(t)$ are plotted with shifts, i.e., $m_x(t) + 3$ and $m_y(t) + 1$ to avoid overlap of curves. The spin spin-length fidelity $s_f(t)$ is also shifted as $s_f(t) - 2$. The field is swept from positive side to the negative side

In this figure, the field is swept from positive side to the negative side. The z components m_z drops around the Stoner-Wohlfarth point which is $H \simeq -1$ in the present case. After this point, m_z continues to have a negative value, but around $H = -10$ it shows a beating. The x and y components, m_y and m_z, respectively, are plotted with a shift $+1$ and $+3$. Here we find a large amplitude of beating. This beating is repeated with a regular period. In the figure, we also plot the quantity

$$s_f(t) = \langle s_x \rangle^2 + \langle s_y \rangle^2 + \langle s_z \rangle^2 \tag{6.144}$$

which we call 'spin-length fidelity'. This quantity changes in time although the total spin is always constant.

$$\langle s_x^2 \rangle + \langle s_y^2 \rangle + \langle s_z^2 \rangle = \frac{S(S+1)}{S^2.} \tag{6.145}$$

We find that the beating is not a rotation of the spin but a beating of spin length.

The period is

$$T = \frac{2\pi S}{D_{\text{SW}}}, \tag{6.146}$$

which is in the classical time scale

$$T_{\text{classical}} = \frac{2\pi S^2}{D_{\text{SW}}}. \tag{6.147}$$

This period comes from an interference of spin oscillation between neighboring energy levels. The energy difference of the level m and $m + 1$ is

$$\Delta_{m,m+1} = E(m+1) - E(m) = \frac{2D_{\rm SW}M + 1 - H}{S}. \tag{6.148}$$

The angular frequency of the interference between the oscillations of $\Delta_{m,m+1}$ and $\Delta_{m+1,m+2}$ is

$$\omega = \Delta_{m,m+1} - \Delta_{m+1,m+2} = \frac{2D_{\rm SW}}{S}, \tag{6.149}$$

which gives (6.146) [113].

In a quantum system, as we see in Fig. 6.24, the population scattered to the kth state remains at the state because no quantum mechanical process changes it. In reality, as we discussed in Sect. 6.7.2, it relaxes to lower energy states by dissipation effects, which will be shown in the next subsection. But, in a fast sweeping, we may see a pure quantum dynamics. Although such velocity may be unrealistic large, the beating is one of characteristics of quantum mechanical system and expected to be found in some experiment in the future.

6.10.2 Distribution of the Population over the States

The wave function is expressed by a linear combination of basis states. Here, we take the eigenstates of the Hamiltonian $\mathcal{H}(t)$ at a given time t as the basis, and express the wave function as

$$\Psi(t) = \sum_{k=1}^{2S+1} c_k|k\rangle, \quad \mathcal{H}(t)|k\rangle = E_k|k\rangle. \tag{6.150}$$

For a large negative field (final stage), the kth eigenstate has the magnetization $m_k = -(S-k)/S$. At the avoided level crossing between the state of S and the kth state, some population is scattered to the kth state. Let the field of the crossing point be $H = H_k$.

The population of the state $P(k) = |c_k|^2$ at the final field is amount scattered from the state of the metastable branch to the state. Thus, $P(k)$ gives information on how the population of the metastable branch was scattered during the sweep. From $P(k)$, we know the distribution of the amount of scattered population as a function of field, and we define $P_{H_k}(H)$ using the k dependence of H_k as

$$P_{H_k}(H)dH = P(k)\left(\frac{dH_k}{dk}\right)dk. \tag{6.151}$$

The peak of $P_{H_k}(H)$ gives the field at which the population scattered mostly. It has been found [115] that when S becomes large, width of the distribution decreases as $S^{-2/3}$, and the mean approaches to $H_{\rm SW}$. It was also found that for nonzero sweeping velocity v, the peak deviates from $H_{\rm SW}$ as

$$|H_{\text{peak}} - H_{\text{SW}}| \propto v^{2/3}. \tag{6.152}$$

This is a shift of the Stoner-Wohlfarth point due to in a finite velocity of sweeping field. This result holds in the limit $S \to \infty$, i.e., in classical case. This shift is difficult to estimate in analysis of the classical system itself.

6.10.3 Dynamics of Magnetization in Dissipative Environment

As we saw in the previous subsection, after H_{SW}, the population distributes over many states. They do not relax to the ground state and show the complicated motion.

In a dissipative environment, the disturbed states relax to the ground state by losing the energy as we saw in Fig. 6.19, and finally it moves to the ground state, and the magnetization relaxes to -1. In Fig. 6.28, we show examples obtained by a master equation with Readfield type dissipation (see Appendix 8.4.2) using $X = s_x$ with the constants $\pi\lambda^2/\hbar = 0.01$ and 0.05. The magnetizations in the dissipative environments move to the equilibrium value $m \simeq -1$. In the case with the strong dissipation $\pi\lambda^2/\hbar = 0.05$, the beating is almost smeared out.

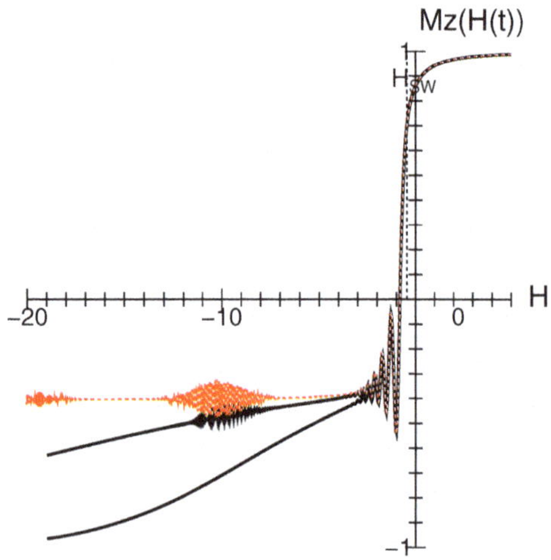

Fig. 6.28 Magnetization curve $m_z(t)$ under sweeping field with the Readfield type dissipation using $X = s_x$ with the constants $\pi\lambda^2/\hbar = 0.01$ and 0.05. $\Gamma = 1$ and $H_x = 1$ and $v = 0.08$. The dotted line shows the magnetization without dissipation (Fig. 6.27)

6.11 Nucleation in Quantum Systems

So far, we studied metastable states in quantum systems for a single variable, e.g., the position of a particle and a spin of S. In the previous sections, we studied a model of large S in which S is conserved. If we consider a system consisting of N spins of $S = 1/2$, the dimension of Hilbert space is 2^N, while the dimension of the subspace of the total spin S is $2S + 1$ ($S = N/2$). To conserve the total spin S during the time evolution, all the operations must be symmetric under exchanges of spins. Thus, the Hamiltonian must be a peculiar property to keep the symmetry, which is realized when all the spins couple with each other. In such systems, the distance between spins has no sense, and we cannot discuss the nucleation, i.e., a local break-down of order.

To study the collapse of metastable state by the nucleation, we have to study systems with short-range interactions. For such systems, we have to study the Hilbert space of 2^N. An attempt to study on metastability in such a system has been done as a quantum spinodal phenomena [116], but it was only in small systems, and more study is necessary to clarify characteristics of the process.

In the next section, we study a characteristics of spin configuration in short-range interacting system by a quantum Monte Carlo simulation with thermal and quantum fluctuation.

6.12 Transverse-Ising Model

In this section, we consider a system exhibiting a phase transition. One of the simplest models which show a quantum phase transition is the transverse Ising model [117][4]

$$\mathcal{H}_{\rm TI} = -J \sum_{<ij>} \sigma_i^z \sigma_j^z - \Gamma \sum_{i=1}^{N} \sigma_i^x, \tag{6.153}$$

where $< ij >$ denotes the sum of all the nearest neighbor pairs. Here we consider the model on lattices in which each site has z nearest sites, e.g., $z = 2$ in a one-dimensional chain and $z = 4$ in a square lattice.

[4] The model was originally given in the form:

$$\mathcal{H}_{\rm TI0} = -J \sum_{<ij>} \sigma_i^x \sigma_j^x - \Gamma \sum_{i=1}^{N} \sigma_i^z$$

which can be transformed to a free fermion system in one-dimensional case, and it is called "transverse Ising model" but not "transverse-field Ising model".

This model exhibits a quantum phase transition at $T = 0$ with respect to the transverse field Γ. The transverse field Γ gives quantum fluctuation on the order of z component of magnetization:

$$M_z = \sum_{i=1}^{N} \sigma_i^z. \tag{6.154}$$

Because σ_i^x does not commute with σ_i^z, it causes a flip of the z component of the spin:

$$\sigma_i^x |\pm\rangle = |\mp\rangle. \tag{6.155}$$

At $T = 0$, when $\Gamma = 0$, the system (6.153) is in the completely ferromagnetic ordered state. For $\Gamma = \infty$, the magnetization aligns along the x axis and the expectation value of M_z becomes zero.

In the corresponding classical model

$$\mathcal{H}_{\text{TIclassical}} = -J \sum_{<ij>} \cos(\theta_i)\cos(\theta_j) - \Gamma \sum_{i=1}^{N} \sin(\theta_i), \tag{6.156}$$

the ground state configuration is given by

$$\frac{d}{d\theta}\left(-\frac{Jz}{2}\cos^2\theta - \Gamma\sin\theta\right) = \frac{Jz}{2}2\cos\theta\sin\theta - \Gamma\cos\theta = \cos\theta\,(Jz\sin\theta - \Gamma) = 0. \tag{6.157}$$

Thus, in the ground state, the x component is given by

$$\sin\theta = \begin{cases} \frac{\Gamma}{Jz} & Jz \geq \Gamma \\ 0 & Jz \leq \Gamma \end{cases}, \tag{6.158}$$

i.e., when $\Gamma \geq zJ$ spins perfectly align along the x direction.

In contrast, in the quantum case, M_z becomes zero even when $M_x/N < 1$ due to the quantum fluctuation. Thus, an order-disorder type phase transition exists as a function of Γ. A schematical phase diagram is depicted in Fig. 6.29. In the one-dimensional system, there is no phase transition at nonzero temperatures ($T > 0$), but at $T = 0$, a phase transition occurs at $\Gamma = J$. At $\Gamma \geq J$, $\langle M_z\rangle = 0$, while at $\Gamma \leq J$, $\langle M_z\rangle$ the spontaneous order appears. In higher dimensions, the spontaneous magnetization appears at a finite critical temperature $T_{\rm C}(\Gamma)$. $T_{\rm C}(\Gamma)$ is reduced by the quantum fluctuation due to Γ, and becomes zero at the quantum phase transition point $\Gamma_{\rm C}$.

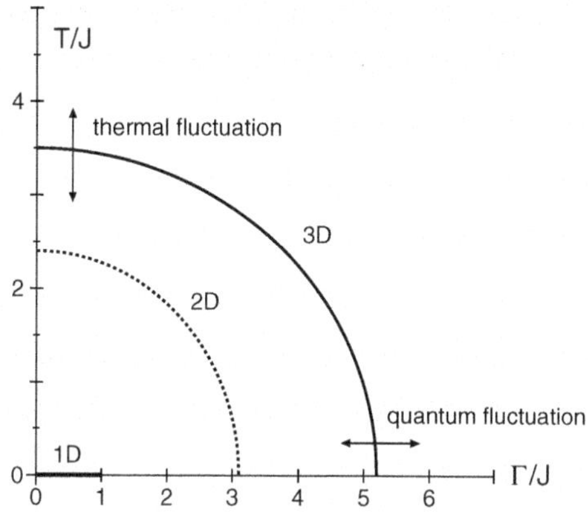

Fig. 6.29 Schematical phase diagram of the transverse Ising model in one (solid line), two (dotted curve), and three (solid curve) dimensions. The one-dimensional model has no ordered state at $T > 0$ and shows a transition only at $T = 0$

6.12.1 Visualization of Quantum and Classical Fluctuation in a (d + 1) Dimensional Representation of States

Here, we make an attempt to visualize the difference between the thermal and quantum fluctuation. For this purpose, we use a path integral representation of the density matrix. The density matrix for equilibrium state at finite temperature is given by

$$\rho = \frac{1}{Z} e^{-\beta \mathcal{H}}, \quad Z = \sum_{\{\sigma_i\}} \langle \{\sigma_i\} | e^{-\beta \mathcal{H}} | \{\sigma_i\} \rangle. \tag{6.159}$$

The idea comes from the quantum Monte Carlo method [118]. The diagonal matrix element of density matrix of a d-dimensional system $\langle \{\sigma_i\} | \rho | \{\sigma_i\} \rangle$ is the probability of the state $|\{\sigma_i\}\rangle$ in the equilibrium state. But it is difficult to calculate the probability, because generally the states $\{\{|\sigma_i\rangle\}$ are not the eigenstates of the Hamiltonian $\mathcal{H}$.

To overcome this difficulty in short-range interacting systems, $\langle \{\sigma_i\} | \rho | \{\sigma_i\} \rangle$ is expressed by a sum of weights of local configuration in a $(d + 1)$-dimensional system by introducing a new dimension called 'imaginary time axis'. This expression is a kind of path integral representation, in which quantum fluctuation is visualized as a fluctuation in the direction of the imaginary time axis.

Let us consider a Hamiltonian consisting of two parts A and B which do not commute. The density matrix is approximated as

$$e^{-\beta(A+B)} \simeq \left(e^{-\frac{\beta}{n} A} e^{-\frac{\beta}{n} B} \right)^n \tag{6.160}$$

for a large n which is called Suzuki-Trotta formula [118].

We assume that operators in A commute with each other, and also in B. For example,

$$A = -J \sum_{<ij>} \sigma_i^z \sigma_j^z, \quad B = -\Gamma \sum_{i=1}^{N} \sigma_i^x. \tag{6.161}$$

The partition function is given by

$$Z \simeq \sum_{\{\sigma_i = \pm 1\}} \langle \{\sigma_i\}| \left(e^{A/n} e^{B/n}\right)^n |\{\sigma_i\}\rangle. \tag{6.162}$$

Substituting complete sets

$$\sum_{\{\sigma_i^m = \pm 1\}} |\{\sigma_i^m\}\rangle\langle\{\sigma_i^m\}|, \tag{6.163}$$

the partition function reads

$$Z = \sum_{\{\sigma_i = \pm 1\}} \sum_{\{\sigma_i^{(1)} = \pm 1\}} \cdots \sum_{\{\sigma_i^{(2n-1)} = \pm 1\}}$$

$$\langle\{\sigma_i\}|e^{A/n}|\{\sigma_i^{(1)}\}\rangle\langle\{\sigma_i^{(1)}\}|e^{B/n}|\{\sigma_i^{(2)}\}\rangle\langle\{\sigma_i^{(2)}\}|\cdots$$

$$\cdots e^{A/n}|\{\sigma_i^{(2n-1)}\}\rangle\langle\{\sigma_i^{(2n-1)}\}|e^{B/n}|\{\sigma_i\}\rangle. \tag{6.164}$$

Here, the partition function is regarded as a partition function of a $(d+1)$-dimensional system in the extended space of $N \times n$ spins: $\{\sigma_i, \sigma_i^{(1)}, \ldots \sigma_i^{(2n-1)}\}$, $(i = 1, \ldots N)$. Here, the suffix n gives an additional dimension for the direction of the temperature. This corresponds to the time axis in the standard path integral representation, and thus it is called 'imaginary time':

$$e^{-i\mathcal{H}t/\hbar} \Leftrightarrow e^{-\beta\mathcal{H}/\hbar} \rightarrow it \Leftrightarrow \beta. \tag{6.165}$$

For the transverse Ising model, using (6.161), the part $\langle\{\sigma_i^{(k)}\}|e^{A/n}|\{\sigma_i^{(k)}\}\rangle$ is given by

$$\langle\{\sigma_i^{(k)}\}|e^{\frac{\beta J}{n}\sigma_i^z\sigma_{i+1}^z}|\{\sigma_i^{(k)}\}\rangle = e^{\frac{\beta J}{n}\sigma_i^{(k)}\sigma_{i+1}^{(k)}}. \tag{6.166}$$

For the part of $\langle\{\sigma_i^{(k)}\}|e^{\tau B}|\{\sigma_i^{(k)}\}\rangle$, we use the relationship:

$$\langle\sigma|e^{\beta\Gamma\sigma_x}|\sigma'\rangle = \sqrt{\frac{1}{2}\sinh(2\beta\Gamma/n)}e^{-\frac{1}{2}\ln(\tanh(\beta\Gamma/n))\sigma\sigma'}, \tag{6.167}$$

and it is expressed as

$$\langle\{\sigma_i^{(k)}\}|e^{\tau\Gamma\sum_{i=1}^{L}\sigma_i^x}|\{\sigma_i^{(k+1)}\}\rangle = \prod_{i=1}^{L}\langle\{\sigma_i^{(k)}\}|e^{\tau\Gamma\sigma_i^x}|\{\sigma_i^{(k+1)}\}\rangle$$

$$= \prod_{i=1}^{L} A e^{K_2\sigma_i^{(k)}\sigma_i^{(k+1)}} = A^L e^{\sum_k^{n-1} K_2\sigma_i^{(k)}\sigma_i^{(k+1)}}, \tag{6.168}$$

where

$$A = \sqrt{\frac{1}{2}\sinh(2\beta\Gamma/n)}, \quad K_2 = -\frac{1}{2}\ln(\tanh(\beta\Gamma/n)). \tag{6.169}$$

Thus, setting

$$K_1 = \frac{\beta J}{n}, \tag{6.170}$$

the model in $(d+1)$-dimensional system is regarded as an Ising model in $(d+1)$ dimensions with the Boltzmann weight:

$$e^{\beta\mathcal{H}_{\text{Ising}}} = e^{K_1\sum_{i,j}\sigma_{i,j}\sigma_{i+1,j}+K_2\sum_{i,j}\sigma_{i,j}\sigma_{i,j+1}}. \tag{6.171}$$

The corresponding lattice in $(d+1)$ dimensions is depicted in Fig. 6.30.

In Fig. 6.31, we show configurations obtained by a quantum Monte Carlo simulation. At a high temperature, K_1 is small and K_2 is large, and thus spins align along the imaginary axis and fluctuate in the real axis with a short-range correlation as shown in Fig. 6.31a. This is standard thermal fluctuation.

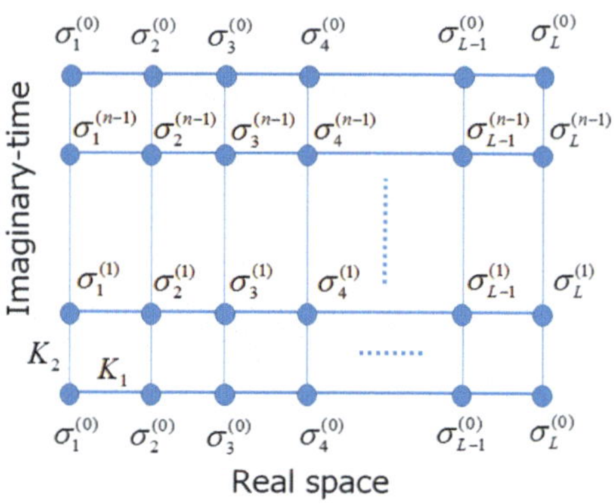

Fig. 6.30 Ising model in two dimensions transformed from the one-dimensional transverse Ising model by the Suzuki–Trotter relation

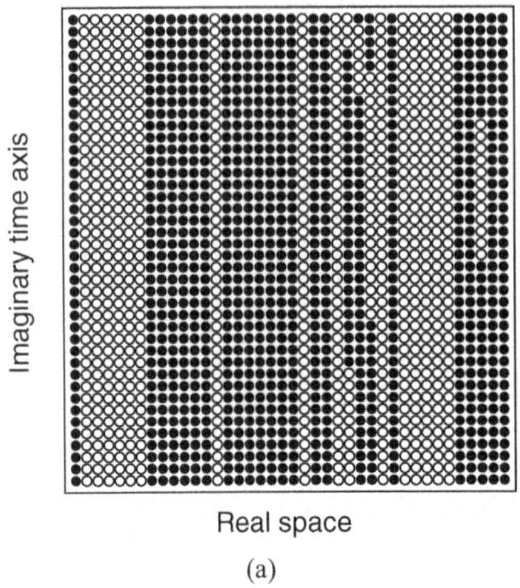

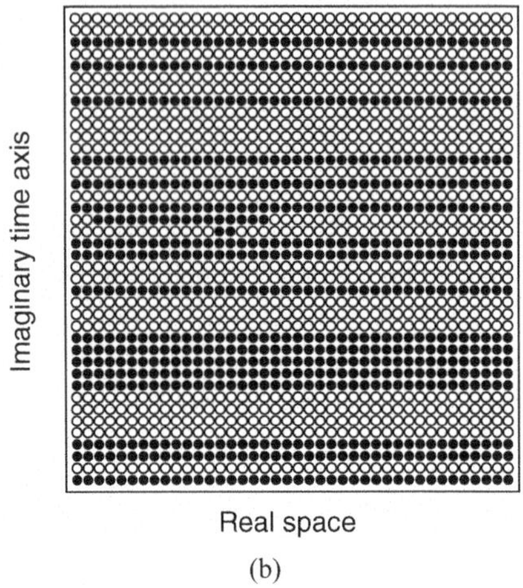

Fig. 6.31 **a** Typical configuration with thermal fluctuation at a high temperature. $T = 2$ and $\Gamma = 0.8$. **b** Typical configuration with quantum fluctuation at a low temperature. $T = 0.01$ and $\Gamma = 2$

On the other hand, at low temperature, K_1 is large and K_2 is small. In this case, spins align along the real axis forming a long correlation. Along the imaginary, the coupling K_2 is small and spins along the real axis flip together as shown in Fig. 6.31b. This is regarded as a quantum tunneling of magnetic clusters.

This quantum fluctuation is controlled by the strength of the transverse field. By making use of this fact, the so-called quantum annealing method was proposed [119] and it has been realized and developing [120].

The one-dimensional transverse Ising model is mapped to two-dimensional Ising model whose free energy has been solved analytically. The critical point of the model in the square lattice with coupling constants for vertical and horizontal directions J_1 and J_2, respectively, is given by

$$\sinh(2\beta_c J_1)\sinh(2\beta_c J_2) = 1. \tag{6.172}$$

Substituting (6.169) and (6.170) to this relation, we find the following relation of J and Γ for the critical point:

$$\sinh(2\beta J/n)\sinh(-\ln(\tanh(\beta\Gamma/n))) = \frac{\sinh(2\beta J/n)}{\sinh(2\beta\Gamma/n)} = 1. \tag{6.173}$$

Thus, the critical transverse field of the one-dimensional model is

$$\Gamma_C(1D) = J. \tag{6.174}$$

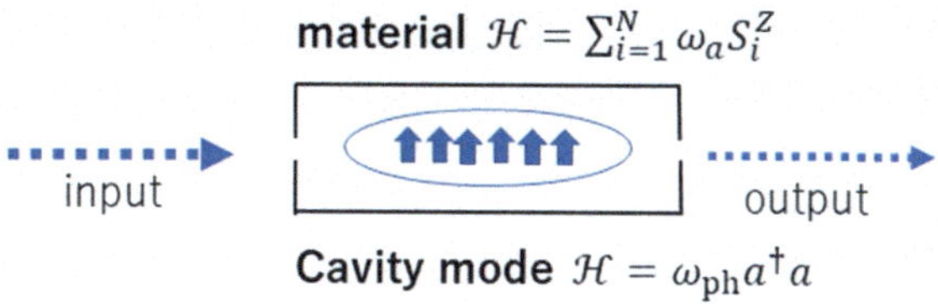

Fig. 6.32 A cavity system consisting of a cavity mode of the eigenfrequency ω_{ph} and ensemble of systems (atom or spin) with discrete energy levels. The photon of the cavity mode is driven by the input and emits the output

6.13 Cooperative Phenomena in a Cavity System

6.13.1 Cavity System

A microcavity has its own resonance frequency ω_{ph} of electromagnetic field (cavity mode), and the field couples with atoms or spins in the cavity as schematically illustrated in Fig. 6.32. Here, we assume that the atom in the cavity has a discrete energy level structure, and they interact with photon of the resonance mode. For simplicity, we assume that the atom is a two-level system, and express the state of atom at the ith site using a spin operator $\boldsymbol{S}_i$ of $S = 1/2$:

$$S_i^z|\pm\rangle = \pm\frac{1}{2}|\pm\rangle, \quad S_i^{\pm}|\pm\rangle = |\mp\rangle, \tag{6.175}$$

where the state $|-\rangle$ is the ground state and $|+\rangle$ is the excited state. In this section, we call S_i 'spin'. The photon is expressed by boson operators a and $a^\dagger$

$$[a, a^\dagger] = 1. \tag{6.176}$$

The input is a classical electromagnetic field driving the cavity mode.

The Hamiltonian of the system is given by the Dicke model [121]:

$$\mathcal{H}_0 = \omega_{\text{ph}} a^\dagger a + \sum_{i=1}^{N} \omega_{\text{a}} S_i^z + i\tilde{g}(a^\dagger - a)\sum_{i=1}^{N}(S_i^+ + S_i^-) + i\tilde{\xi}(a^\dagger e^{-i\Omega t} - ae^{-i\Omega t}), \tag{6.177}$$

where ω_{ph} is the resonance frequency of the cavity, and ω_{a} is the energy gap of the atom. The interaction between photon and an atom is given by a dipole approximation with the coefficient $\tilde{g}$. The last term is the external driving force representing the input field with the circular polarization whose amplitude is given by $\tilde{\xi}$:

$$i(a^\dagger - a)\cos\Omega t + (a^\dagger + a)\sin\Omega t = i\left(a^\dagger e^{-i\Omega t} - ae^{-i\Omega t}\right). \tag{6.178}$$

The model with the rotating wave approximation is called Jaynes–Cummings model [122] (single atom), or Tavis-Cummings model (many atoms) [123].

$$\mathcal{H}_0 = \omega_{\rm ph} a^\dagger a + \sum_{i=1}^{N} \omega_{\rm a} S_i^z - i\tilde{g} \sum_{i=1}^{N} (a S_i^+ - a^\dagger S_i^-) + i\tilde{\xi}(a^\dagger e^{-i\Omega t} - a e^{-i\Omega t}), \tag{6.179}$$

where the terms $a^\dagger S_i^+$ and $a S_i^-$ are discarded.[5] These models have been used to analyze properties of the so-called cavity quantum electrodynamics (cavity QED)[124].

Hereafter, we consider the Tavis-Cummings model. In the following, we mainly study the case of the resonance, i.e.,

$$\omega_{\rm a} = \omega_{\rm ph} = \Omega = \omega. \tag{6.181}$$

The state is specified by the number of photons n and the magnetization M

$$n = a^\dagger a, \quad M = \sum_{i=1}^{N} S_i^z, \quad S_i = \pm 1/2 \tag{6.182}$$

and we may express the state by $|n, M\rangle$. However, when the system has N_+ up spins, i.e., $N_- = N - N_+$ down spins, the magnetization is $M = -N/2 + N_+$. Thus, hereafter, we denote the state by $|n, N_+\rangle$ instead of $|n, M\rangle$.

When $\tilde{g} = 0$, the ground state is $|0, 0\rangle$ with the energy $E_0 = -N\omega/2$ and the first excited states are $|1, 0\rangle$ and $|0, 1\rangle$ with the energy $E_1 = E_0 + \omega$. The magnetic resonance mentioned in the footnote corresponds to the transition between them: $(|0, 0\rangle \leftrightarrow |1, 0\rangle$ or $|0, 1\rangle)$. When $\tilde{g} \neq 0$, the states $|1, 0\rangle$ and $|0, 1\rangle$ hybridize and an energy splitting appears with the energy gap

$$\Delta E = 2\sqrt{2S}\tilde{g}, \quad S = N/2, \tag{6.183}$$

which was pointed out by Agarwal [126] and the splitting is called 'vacuum Rabi splitting'. When this type of coupling is relevant, the system is called to be in 'strong coupling regime'. It has been found that when the number of photons is smaller than N, the energy levels are strongly modified by the effects of the coupling. When the number of photons is large, the photon behaves as a classical electromagnetic field ($a \to He^{-i\Omega t}$). There, the energy levels are almost equally separated, and the feature of classical magnetic resonance is recovered. Thus, between the two cases, qualitative different features appear [127, 128]. In the right panel of Fig. 6.33, we

[5] If we treat the electromagnetic field classically, i.e., $a \to He^{-i\Omega t}$, the present model is reduced to a model of magnetic resonance, e.g. ESR or NMR [125],

$$\mathcal{H}_{\rm ESR} = \sum_{i=1}^{N} \omega_{\rm a} S_i^z - 2H \sum_{i=1}^{N} (S_i^x \sin(\Omega t) - S_i^y \cos(\Omega t)). \tag{6.180}$$

.

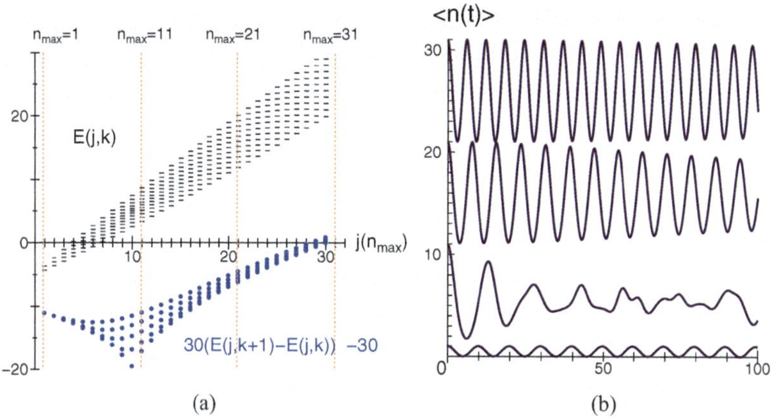

Fig. 6.33 **a** The energy levels in each block specified by the photon number $j(= n_{\max})$ are plotted by bars. The energy differences between neighboring levels (6.185) are plotted by dots, where the date are multiplied by 30 and shifted by -30 to make them visible. **b** Time evolution of photon number starting with the maximum value in each block for $j = 1, 11, 21$ and 31

show the eigenenergy levels for $N = 10$ with $\tilde{g} = 0.1$ by short bars. Because the Hamiltonian is block diagonalized with the parameter

$$j = n + N_+, \tag{6.184}$$

the energy levels in each block $E_j(k)$ are plotted separately. The ground state $|0, 0\rangle$ is a state of $n = 0$ and $N_+ = 0$, i.e., $j = 0$, and the energy is $E_0(1) = 0$. When one photon is added to the ground state, the states $|1, 0\rangle$ and $|0, 1\rangle$ appear, and they form the block of $j = 1$. These states hybridize and two eigenstates appear with the eigenvalues $E_1(1)$ and $E_1(2)$. In this way, by adding photons, the number of states in block increases until the number of added photons becomes N. There, $j \leq N$, and there are $j + 1$ eigenstates with $E_j(k), k = 1, \ldots j + 1$. When the number of added photons becomes larger than N, the number of states is fixed to be $N + 1$ because there are only $N + 1$ states, $|j, 0\rangle$,$|j - 1, 1\rangle, \ldots |j - N, N\rangle$. The eigenvalues are $E_j(k), k = 1, \ldots, N + 1$ for $j \geq N$.

The energy gaps between neighboring levels in each block j, i.e.,

$$\Delta E_j(k) = E_j(k + 1) - E_j(k) \tag{6.185}$$

are plotted by dots below the figure, where the values are multiplied by 30 and shifted by -30 to make them visible. The vacuum Rabi splitting mentioned above is $\Delta E_1(1)$.

The left panel shows the time dependence of the photon number starting from $n(0) = j$ and $N_+ = 0$. For $j = 1$, we find a simple oscillation with the frequency given by the vacuum Rabi splitting. For large j, we find again almost single mode oscillation whose frequency is that of the Rabi oscillation of the single spin case.

Around $j = N$, we find a kind of phase transition where the energy gaps $\{\Delta E_j(k)\}$ are largely distributed and $n(t)$ shows a decoherent behavior.

The magnetic resonance is a process changing j by 1. The energy difference is

$$\Delta_{j,k,l} = E_{j+1}(l) - E_j(k), \tag{6.186}$$

which gives the resonance frequency

$$\omega_{\mathrm{Res}} = \Delta_{j,k,l}. \tag{6.187}$$

The amplitude $X_{k,i,j}$ of the resonance is given by the matrix element of the operator a as

$$X_{k,i,j} = |\langle n, m(n+m=j)|a|n', m'(n'+m'=j+1)\rangle|^2. \tag{6.188}$$

It was found that $X_{j,k,l}$ is nearly zero except $|n-n'| = 1$. In the region of $n < N$, the resonance frequencies with large amplitudes locate near those between the ground state $|0, 0\rangle$ and the vacuum Rabi splitting (i.e., $E_1(1)$ and $E_1(2)$), which give two peaks of resonance. In contrast, for large j, the splitting (6.186) locates around the resonance frequency in the classical magnetic resonance, i.e., $\omega_{\mathrm{Res}} = \omega_a$. Thus, there are two regions depending on the number of photon in the cavity [128]. The splitting of resonance frequency for small photon density (i.e., small j) explains the observation of splitting of ESR spectrum of the paramagnetic material DPPI [127]. Similar spectrums with splitting have been observed in various materials in the strong coupling region [129].

6.13.2 *Phase Transitions of the Dicke Hamiltonian*

There are several kinds of phase transitions that occur in the cavity.

6.13.2.1 Dicke Transition

As a phase transition in the cavity system, the photon number in the ground state changes from 0 to a nonzero value at the strength of the coupling $\tilde{g}_{\mathrm{C}} = \sqrt{\omega_{\mathrm{a}}\omega}/\sqrt{N}$, which is called 'Dicke transition'. When $\tilde{g} > \tilde{g}_{\mathrm{C}}$, photon spontaneously appears without driving force [130, 131]. As to this transition, there are arguments about the effect of second-order contribution of the electromagnetic field which inhibits the phase transition (no-go-theorem) [132].

6.13.2.2 Phase Transition in a Driven System

When the present model is in a dissipative environment, the system reaches a stationary state in a balance of the driving force and the dissipation. The number of photon in the cavity in the stationary state shows a phase transition as a function of the strength of the driving field $\tilde{\xi}$, which causes a discrete change of the transmission through the cavity. This phase transition is related to the optical bistability, which will be studied in detail in the next section.

Besides the optical bistability, rich structures of phase transitions have been found in region of strong coupling and strong driving force [131]. In the strong coupling region, the Dicke model and the Tavis-Cummings model have qualitatively different phase diagrams. For details, see the original paper [131].

6.14 Optical Bistability

In this section, we study a first-order phase transition in a dynamical process. As an example, we study a phenomenon called 'optical bistability'. When the light transmits in a medium or in the cavity with a gas, the light interacts with atoms (or spins) with a discrete energy level. The ratio of amplitudes of the input and output is the transmission constant. In some materials, the transmission constant changes discontinuously with the strength of input. This phenomenon is owing to the interaction between the light and material, and it has been studied as a problem of classical optics [133, 134]. Here, we adopt the Dicke model (6.177) to study this phenomenon quantum mechanically [135].

We study case of the resonance condition, i.e.,

$$\omega_{\mathrm{ph}} = \omega_{\mathrm{a}} = \Omega = \omega. \tag{6.189}$$

We consider cases of high frequency, and use the rotating wave approximation

$$(a^\dagger - a)(S_i^+ + S_i^-) \to a^\dagger S_i^- - aS_i^+, \tag{6.190}$$

which causes no serious problem to study the optical bistability phenomena although it gives serious difference in strong driven systems [131].

The Hamiltonian with the rotating wave approximation is

$$\mathcal{H}_{\mathrm{R}} = \omega_{\mathrm{ph}} a^\dagger a + \sum_{i=1}^{N} \omega_{\mathrm{a}} S_i^z + i\tilde{g}\sum_{i=1}^{N}(a^\dagger S_i^- - aS_i^+) + i\tilde{\xi}(a^\dagger e^{-i\Omega t} - ae^{-i\Omega t}). \tag{6.191}$$

In this section, we put $\hbar = 1$. In this form of Hamiltonian, we can remove the explicit time dependence by introducing a rotational frame with the unitary transformation:

$$U(t) = \exp\left[-i\omega t\left(a^\dagger a + \sum_{i=1}^{N} S_i^s\right)\right]. \tag{6.192}$$

By this transformation, the Hamiltonian reads

$$\mathcal{H}_R \equiv U^{-1}(t)\left(\mathcal{H} - i\frac{\partial}{\partial t}\right)U(t) = i\tilde{g}\sum_{i=1}^{N}(a^\dagger S_i^- - aS_i^+) + i\tilde{\xi}(a^\dagger - a). \tag{6.193}$$

In the present study, we adopt a simple Lindblad type of dissipation term $D[\rho_R]$, and then the quantum master equation is given by

$$\frac{d}{dt}\rho_R = -i[\mathcal{H}_R, \rho_R] + D[\rho_R] \equiv \mathcal{L}\rho_R, \tag{6.194}$$

where

$$D[\rho_R] = D_{\text{photon}}[\rho_R] + D_{\text{atom}}[\rho_R] \tag{6.195}$$

with

$$D_{\text{photon}}[\rho_R] = \kappa[2a\rho_R a^\dagger - (a^\dagger a\rho_R + \rho_R a^\dagger a)] \tag{6.196}$$

and

$$D_{\text{atom}}[\rho_R] = \gamma\sum_{i=1}^{N}\left[2S_i^-\rho_R S_i^+ - (S_i^+S_i^-\rho_R + \rho_R S_i^+S_i^-)\right]. \tag{6.197}$$

These dissipation terms merely represent a decay of photon number and the number of excited atoms, respectively, and do not generally bring the system to the ground state of the given Hamiltonian. To realize the equilibrium state as the stational state of the master equation, we need to use a Redfield-type master equation which is explained in Sect. 8.4.2. However, the choice of (6.196) and (6.197) is good enough to have rough idea of the dissipation effect. Moreover, the present system is driven by a time dependent field, and the state is out of equilibrium. Thus, precise treatment of the dissipation effect is difficult, and we work with the present simple choice of dissipation.

6.14.1 Mean-Field Analysis

We assume that there are no direct interactions among atoms, but atoms interact with each other via the photon in the cavity. In this case, interactions among atoms are independent of the sites of atom, and all the atoms interact with all the atoms. Thus, the interaction is of the infinite-range, for which the mean-field approximation gives correct results in the limit $N \to \infty$ [136].

Therefore, we first study the model in this limit. To have natural limit of quantities in the limit $N \to \infty$, we scale quantities as

$$\alpha(t) = \frac{\mathrm{Tr} a \rho_{\mathrm{R}}(T)}{\sqrt{N}}, \quad m^{\pm} = \mathrm{Tr} S_i^{\pm} \rho_{\mathrm{R}}(T), \quad m^z = \mathrm{Tr} S_i^z \rho_{\mathrm{R}}(T), \tag{6.198}$$

and scale the parameters as

$$g = \sqrt{N}\tilde{g}, \quad \text{and} \quad \xi = \frac{\tilde{\xi}}{\sqrt{N}}. \tag{6.199}$$

Then, the equations of motion of quantities are given as

$$\begin{cases} \dfrac{\partial \alpha}{\partial t} &= (gm^- + \xi) - \kappa\alpha, \\ \dfrac{\partial m^-}{\partial t} &= 2g\alpha m^z - \gamma m^- \\ \dfrac{\partial^z}{\partial t} &= -g(\alpha^* m^- + \alpha m^+) - \gamma(2m^z + 1). \end{cases} \tag{6.200}$$

In the steady state of the equations of motion, the following relation holds.

$$\xi = \left(\kappa + \frac{\gamma g^2}{2g^2\alpha^2 + \gamma^2}\right)\alpha. \tag{6.201}$$

By the scaled parameters

$$\alpha_{\mathrm{s}} = \frac{\sqrt{2}g}{\gamma}\alpha, \quad \xi_{\mathrm{s}} = \frac{\sqrt{2}g}{\kappa\gamma}\xi, \quad C = \frac{g^2}{2\kappa\gamma}, \tag{6.202}$$

(6.201) is rewritten as

$$\xi_{\mathrm{s}} = \left(1 + \frac{2C}{\alpha_{\mathrm{s}}^2 + 1}\right)\alpha_{\mathrm{s}}. \tag{6.203}$$

Here C is called 'cavity cooperative parameter' [137]. The photon density is $n/N = \alpha^2$, which depends on γ and κ as

$$\alpha^2 = \frac{\gamma}{4\kappa C}\alpha_{\mathrm{c}}^2. \tag{6.204}$$

The relation (6.203) indicates that α is a nonmonotonic function of ξ, and there is a region of bistability when $C > 4$. The interval of the bistable region is given by $\xi_- < \xi < \xi_+$:

$$\xi_{\pm} = \frac{\kappa\gamma}{2g}\sqrt{(C^2 + 10C - 2) \pm (C-4)\sqrt{C(C-4)}}. \tag{6.205}$$

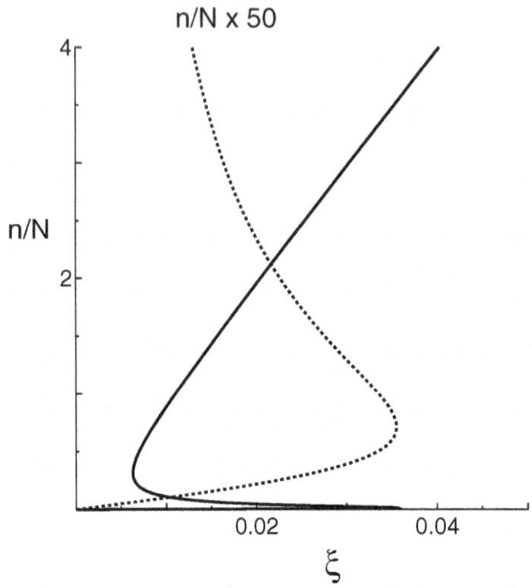

Fig. 6.34 Photon density $n/N = \alpha^2/N$ as a function of the input ξ for $g = 0.1, \gamma = 0.002$. The dotted line shows a magnification ($\times 50$) of the low photon-density part

In Fig. 6.34, we plot the photon number $n = \langle a^\dagger a \rangle = \alpha^2$ as a function of ξ for $g = 0.1$ and $\kappa = 0.05$ and $\gamma = 0.002$. In this case, $C = 50$, and $\xi_- = 1.41 \times 10^{-2}$ and $\xi_+ = 3.61 \times 10^{-2}$. In the region of weak input (small ξ), the photon number is proportional to ξ with a small coefficient, which looks almost flat. To see the value well, the value is multiplied by a factor 50 and depicted by the dotted line in the figure. In the region of strong input (large ξ), the photon number is also nearly proportional to ξ with a larger slope. The photon number does not change smoothly and shows a multivalued structure. In the multivalued region, the system is bistable, and the photon number must jump between the two branches.

This bistable structure is understood in the following picture. The photon number in the cavity is enlarged by the driving force, while the photon number is reduced by the dissipation (6.196). Besides (6.196), the photon number is also reduced through the relaxation of magnetization. Namely, by the interaction between photon and spins, photon excites atoms, and the excited spins relax to the ground state by the dissipation (6.197). Thus, there are two ways of reduction of photon number. When the number of photons is small, both dissipation mechanisms work for all the photons. On the other hand, in the strongly driven case (large ξ) the photon number is large, and the relaxation via $D_{\text{atom}}[\rho_{\text{R}}]$ is saturated and only $D_{\text{photon}}[\rho_{\text{R}}]$ gives a dominant role. Because of the interaction among spins, a cooperative phenomenon takes place, and a kind of first phase transition occurs, that is the optical bistability.

6.14.2 Analogy to a Picture of Thermodynamic Free Energy

The dependence of n on ξ has the same singular structure as that of a first-order phase transition. We can determine the positions of the singular points, i.e., the spinodal points, by (6.205). But, we cannot obtain the first-order phase transition point. In equilibrium systems, the first-order phase transition point is given as a point where the free energies per a degree of freedom $f(\alpha)$ of both phases, say A and B, are the same:

$$f(\alpha_{\rm A}) = f(\alpha_{\rm B}), \tag{6.206}$$

where α is a parameter which controls the state, e.g., the temperature T and the magnetic field H. The idea of Maxwell's equal area law is used to determine it. However, in dynamical systems, the free energy does not exist. In this situation, we may define the critical point as a point where the distributions of population in the limit $N \to \infty$ are the same:

$$P(\alpha_{\rm A}) = P(\alpha_{\rm B}). \tag{6.207}$$

In equilibrium systems, the probability and the free energy are related as

$$P(\alpha) \propto e^{-\beta N f(\alpha)}, \tag{6.208}$$

and the both definitions (6.206) and (6.207) are the same. In the present case, β has no meaning and just a constant parameter. Although, we do not have the free energy, we can still use (6.207). Now, we determine the position of the transition point from size dependence of probability distribution $P(\alpha)$.

In the picture of the first-order phase transition in the thermodynamic systems where the free energy gives the probability as (6.208), the difference of the free energies increases proportionally to the size of the system, which causes the difference of the probabilities increase exponentially.

Making use of an analogy to this picture, we study the critical point of the optical bistability. At the first-order phase transition point, the probabilities of both phases are the same, and the averaged value of the quantity, e.g., the photon number, is given by the average of values in the two phases, while if the system moves out from the transition point, one of the phases becomes relevant and the probabilities of the phases become different. As the size of the system increases, the difference increases. Thus, the averaged value moves to the value of the stable phase. In the limit $N \to \infty$, the value reaches the value of stable state. Making use of this size dependence, we determine the first-order phase transition point. We may call the transition as 'first-order phase transition', if the ratio, $P(\alpha_{\rm A})/P(\alpha_{\rm B})$, changes exponentially with the number of spins N. We call this way to determine the critical point 'free energy picture'.

Although we do not have the "free energy" in dynamical systems, one might still try to obtain free energy by assuming the van Hove type equation, i.e.,

$$\frac{dn}{dt} = -\frac{1}{\tau}\frac{dF(n)}{dn} \tag{6.209}$$

because we know dn/dt. But this idea does not work, because τ may depend on n, and actually we can check that it gives a wrong result. If we need 'free energy' of the present model, we could be determine it by the relation (6.208) as

$$f(\alpha) = -\frac{1}{N}\frac{1}{\beta}\ln P(\alpha). \tag{6.210}$$

6.14.3 Numerical Study of the Size Dependence

We first check whether the free energy picture holds or not. That is, how the probability distribution of n changes with N. In the free energy picture, we expect the following dependence on N. When ξ is larger than the critical value ξ_c, the probability of the upper branch increases when N increases. On the other hand, when ξ is smaller than the critical value ξ_c, the probability of the lower branch increases. Thus, when N increases, $\langle n\rangle$ moves in opposite directions when ξ is below or above ξ_c. Thus, the slope of $\langle n\rangle$ at the transition point becomes large with N, and n jumps at the point in the limit $N \to \infty$.

Dependence of the average photon number on the size was studied by a large-scale numerical study of quantum master equation [135]. Instead of performing a direct simulation of the dynamics, the stationary state was obtained by solving the equation for the eigenstate of zero eigenvalue of the time-evolution operator $\mathcal{L}$:

$$\mathcal{L}\rho_{ss} = 0, \tag{6.211}$$

where ρ_{ss} is a matrix whose dimension is

$$M = 2^N \times n_{\max}. \tag{6.212}$$

To solve the eigenvalue problem (6.211), we express the elements of ρ_R in a vector representation:

$$\mathcal{L}\rho_i = \lambda_i\rho_i \to \hat{L}\text{æ} = \lambda_i\text{æ}, \tag{6.213}$$

where

$$\text{æ} = \begin{pmatrix} (\rho_R)_{11} \\ (\rho_R)_{12} \\ \vdots \\ (\rho_R)_{MM} \end{pmatrix}. \tag{6.214}$$

In principle $n_{\max} = \infty$, but we truncate it by $2N$. In the case of the infinite-range model, we can use a symmetric property and the dimension for the part of atoms is reduced as

$$2^N \to \frac{(N+2)(N+1)N}{6}, \tag{6.215}$$

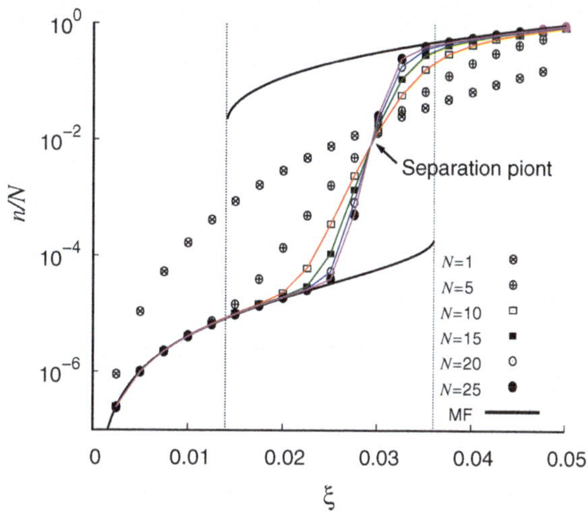

Fig. 6.35 Size dependence of the photon number n in the logarithmic scale as a function of the input ξ obtained by numerical calculation [135]. The bold curves denote the solutions obtained by the mean-field analysis. The dotted lines denote end points of bistability (6.205)

which helps numerical calculation [138–140]. We use a numerical method to obtain the eigenstate of zero eigenvalue. In practice, we obtained non-trivial solution of (6.211). Because $\mathcal{L}$ is not a symmetric matrix, the so-called biconjugate gradient (Bi-CG) method [141] was used.

The obtained photon numbers n as a function of ξ in finite size systems are plotted in Fig. 6.35. There, we find that the slope becomes sharper as N increases. Moreover, we find the curves cross almost at a point noted by 'separation point', which should be a sign of the first-order phase transition point. In the right side of the crossing point, the upper branch is stable state in the thermodynamic limit, while on the left side, the lower branch is stable. This dependence is consistent with the free energy picture.

6.14.4 Metastability in the Bistable Region

As another characteristic of the first-order phase transition, the relaxation time of metastable state increases exponentially with the size of the system. To confirm this property, we studied size dependence of the relaxation time. With eigenstates of the time-evolution operator $\mathcal{L}$, the dynamics of $\rho_{\rm R}$ is given by

$$\rho_{\rm R}(t) = \rho_{\rm ss} + \sum_{i=2}^{M} c_i e^{\lambda_i t} \rho_i, \quad \mathcal{L}\rho_i = \lambda_i \rho_i. \tag{6.216}$$

Here, the eigenvalues are sorted as

$$0 > {\rm Re}\lambda_2 \geq {\rm Re}\lambda_3 \ldots. \tag{6.217}$$

Relaxation time is calculated from the second eigenvalue of the time-evolution operator $\mathcal{L}$

$$\tau = -\frac{1}{\mathrm{Re}\lambda_2}, \quad \lambda_1 = 0. \tag{6.218}$$

The second eigenvalue is estimated by the inverse power method. That is, starting from an initial state x_0, we extract the component of ρ_{ss} as

$$x_1 = x_0 - (\mathrm{Tr}x_0)\rho_{ss}, \tag{6.219}$$

where x_1 consists of eigenstates $\rho_i, i > 1$. Solving x_{k+1} iteratively as a solution of the equation in the vector representation

$$\mathcal{L}x_{k+1} = x_k \tag{6.220}$$

by the Bi-CG method, the relaxation time is obtained as

$$\tau = \lim_{k\to\infty} \left(\frac{||x_k||_{\mathrm{F}}}{||x_{k+1}||_{\mathrm{F}}} \right)^{-1}, \tag{6.221}$$

where $|| \cdot ||$ denotes the Frobenius norm [142].

Thus obtained relaxation times for finite size systems are depicted in logarithmic scale in Fig. 6.36a. In the bistable region $\xi_- < \xi < \xi_+$, the relaxation time increases with the same increment at each ξ for the systems of $N = 10$, 15, 20, and 25, which indicates that the relaxation time increases as

$$\tau \sim e^{c(\xi)N}, \quad c(\xi) > 0. \tag{6.222}$$

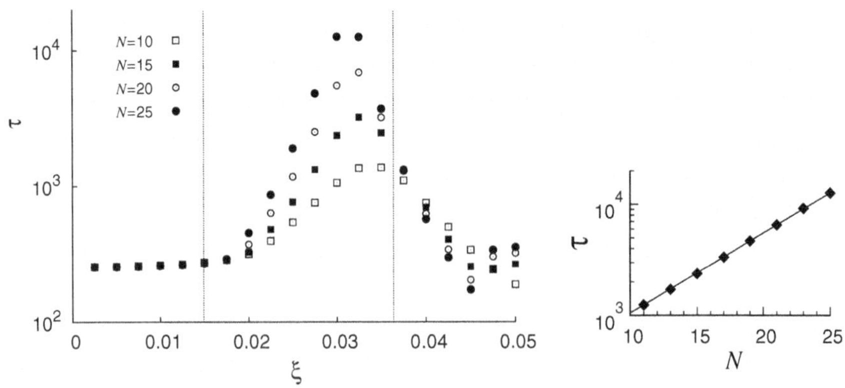

Fig. 6.36 **a** ξ dependence of the relaxation time for various sizes [135]. The dotted lines denote (6.205). **b** Size dependence of the relaxation time at $\xi = 0.03$

There, the constant $c(\xi)$ has the maximum value at the transition point $\xi_c \simeq 0.03$ as

$$c(\xi_c) \simeq 0.166. \tag{6.223}$$

The size dependence of the relaxation time at $\xi = 0.03$ is plotted in Fig. 6.36b. This observation confirms the free energy picture of first-order phase transition.

6.14.5 Hysteresis Phenomena

Here, it should be noted that in the present calculation we study the steady state by obtaining the eigenstate of the time-evolution operator. Thus, the eigenstate gives a probability distribution averaged over infinite time, and the effect of time correlation of state is not taken into account. If we study the state by simulating the dynamics using a time-evolution operator $\mathcal{L}$, the state will show a history dependence. For example, the state starting from the right side tends to stay in the same branch as a metastable state even after the first-order phase transition point.

Because of the large relaxation time, the system shows a hysteresis when we sweep and sweep back ξ over the bistable region. In Fig. 6.37, we show an example of the hysteresis for the system of $N = 25$. There, the sweeping velocity is $v = 2 \times 10^{-7}$. The width of the hysteresis depends linearly on the sweeping velocity and depends on N exponentially. These dependences are typical characteristics of the first-order phase transition [135].

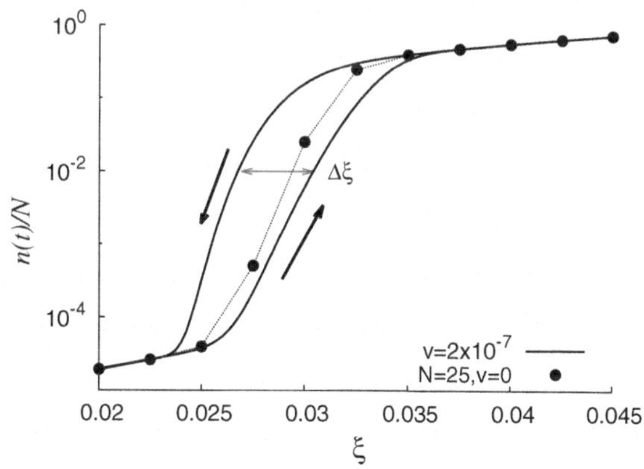

Fig. 6.37 The hysteresis loop of photon number density for $N = 25$ [135]. Dots show the values in the steady state ($v = 0$) which are connected by a dotted line, and the bold lines give $n(t)$ obtained in sweeping processes with $v = \pm 2 \times 10^{-7}$

6.15 Limit Cycle of the Hysteresis

6.15.1 Dynamics Under an Driving Force with Periodically Oscillating Amplitude

In the previous section, we found hysteresis in the field process of sweep and sweep back. The width of hysteresis depends on the sweeping velocity (not shown). Now, it is interesting to study how the magnetization curve changes under a driving force $\xi(t)$ whose amplitude changes sinusoidally with a period T:

$$\xi(t) = h_0 + h_{\text{ex}} \sin\left(\omega_\xi t\right), \quad \omega_\xi = \frac{2\pi}{T}, \tag{6.224}$$

i.e., the driving field change as

$$\mathcal{H}_{\text{drive}} = i\xi(t)(a^\dagger e^{-i\Omega t} - ae^{i\Omega t}). \tag{6.225}$$

Under this driving force with an oscillating amplitude $\xi(t)$, the system repeats hysteresis type motion, and after many oscillations of $\xi(t)$, the magnetization is expected to converge to a limit cycle [143].

Here, we assume that Ω is much larger than ω_ξ. In the rotation frame, the oscillation of the frequency of Ω can be removed as we did in the previous subsection. However, the time dependence of $\xi(t)$ causes time dependence of ρ_{R} even in the rotation frame.

To see how the limit cycle of ρ_{R} changes with ω_ξ, here we study a simplified model. Namely, we look at only dynamics of the atoms by eliminating the photon degree of freedom in the adiabatic approximation [138, 144]. In the model thus obtained, we study the dynamics of density matrix of the atomic degree of freedom $\rho_{\text{a}}(= \text{Tr}_{\text{photon}}\rho_{\text{R}})$ given by (see the detail in the original paper [143])

$$\frac{d\rho_{\text{a}}}{d(\gamma t)} = -iA(t)\sum_{i=1}^{N}\left[S_i^x, \rho_{\text{a}}\right] + \mathcal{D}\rho_{\text{a}}, \tag{6.226}$$

where

$$A(t) = \frac{2g\xi(t)}{\kappa\gamma} = A_0 + A_{\text{eq}} \sin\left(\frac{2\pi}{T}t\right). \tag{6.227}$$

Thus, the amplitude of external field oscillates between $A_0 + A_{\text{eq}}$ and $A_0 - A_{\text{eq}}$. We adopt the dissipation term in a simple Lindblad type of dissipation term as we did in previous sections.

$$\mathcal{D}\rho_{\text{a}} = \frac{2C}{N}\sum_{i=1}^{N}\sum_{j=1}^{N}\left(2S_i^-\rho_{\text{a}}S_j^+ - \{S_i^+S_j^-, \rho_{\text{a}}\}\right) + \sum_{i=1}^{N}\left(2S_i^-\rho_{\text{a}}S_i^+ - \{S_i^+S_i^-, \rho_{\text{a}}\}\right), \tag{6.228}$$

where C is the cooperative parameter [137]

$$C = \frac{Ng^2}{2\kappa\gamma}. \tag{6.229}$$

6.15.2 Floquet Map

To study dynamics, instead of solving the equation of motion directly, we use the Floquet map. Because $A(t)$ changes periodically (6.227), the Hamiltonian changes as

$$\mathcal{H}(t) = \mathcal{H}(t+T), \tag{6.230}$$

and the time-evolution operator has the same period

$$\mathcal{L}(t) = \mathcal{L}(t+T). \tag{6.231}$$

The Floquet map $\mathcal{F}$ is defined by

$$\mathcal{F} \equiv \exp_{\rightarrow}\left(\int_0^T \mathcal{L}(t)dt\right), \tag{6.232}$$

where $\exp_{\rightarrow}$ denotes the time-ordering operator. This operation is called 'Floquet Liouvillian operator'.

The eigenvalues of $\mathcal{F}$ are ordered as

$$1 = \lambda_0 \geq |\lambda_1| \geq \ldots. \tag{6.233}$$

The eigenstate of λ_0 gives the limit cycle

$$\mathcal{F}\rho_0 = \rho_0. \tag{6.234}$$

The trajectory of an operator O in the limit cycle is given by

$$O_{\mathrm{LS}}(t) = \frac{1}{\mathrm{Tr}\rho_0}\mathrm{Tr}\left[O \exp_{\rightarrow}\left(\int_0^t \mathcal{L}(t)dt\right)\rho_0\right]. \tag{6.235}$$

The decay rate to the limit cycle is given by

$$\frac{1}{\tau} = -\frac{1}{T}\mathrm{Re}(\ln \lambda_1). \tag{6.236}$$

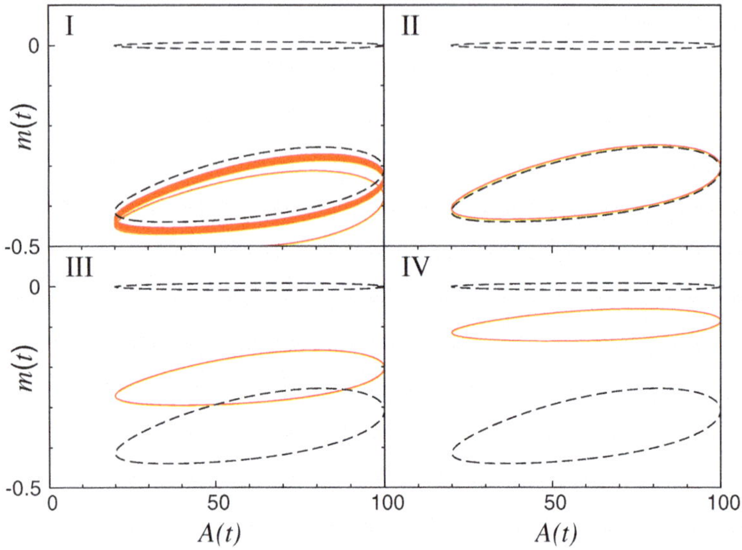

Fig. 6.38 Trajectories of $m_z(t)$ for $N = 70$ and $T = 0.06\gamma^{-1}$ in time intervals: I ($0 \leq t \leq 10T$), II ($100T \leq t \leq 110T$), III ($1000T \leq t \leq 4010T$), and IV finial limit cycle ($t = \infty$). The limit cycles obtained by the mean-field analysis are shown by dashed lines. From the paper [143]

A typical example of the change of $m(t) = \langle \sum_i^N S_i^z \rangle / N$ is depicted in Fig. 6.38. The early stage is shown in I ($0 \leq t \leq 10T$), where starting an initial point ($m(0) = -0.5$, $A(t) = 50$), the trajectory moves to one of the limit cycles obtained by the mean-filed analysis shown by the dashed lines. There are two separate limit cycles obtained by a mean-field analysis which will be explained in the next subsection. Then the trajectory almost converges to one of the limit cycles obtained by mean-filed analysis as given as interval II ($100T \leq t \leq 110T$). However, after a long time, some population leaks out to the other stable state. Thus, the trajectory moves towards to the middle of the mean-filed limit cycles as shown in III ($1000T \leq t \leq 4010T$). Because size of the system is finite $N = 70$, in the final stage, the population exists in both stable branches and the limit cycle numerically obtained in the finite size system is given by a superposition of them, which is shown in IV. This type of change of limit cycle has been studied in the kinetic Ising model [145].

6.15.3 Mean-Field Analysis of Limit Cycle

In this section, we study the time-evolution (6.226) by a mean-field analysis, which is good for $N \to \infty$. The evolution is given by

$$\begin{aligned}
\frac{dm_x}{d(\gamma t)} &= (4Cm_z - 1)m_x, \\
\frac{dm_y}{d(\gamma t)} &= -A(t)m_z + (4Cm_z - 1)m_y, \\
\frac{dm_z}{d(\gamma t)} &= A(t)m_y - 4C(m_x^2 + m_y^2) - (2m_z + 1)(4Cm_z - 1)m_x,
\end{aligned} \tag{6.237}$$

where

$$m_\alpha(t) = \mathrm{Tr} m_\alpha \rho_{\mathrm{MF}}(t), \quad \alpha = x, y, \text{and } z, \tag{6.238}$$

and as we mentioned in (6.229) and (6.227):

$$C = \frac{Ng^2}{2\kappa\gamma}, \quad A(t) = \frac{2g}{\kappa\gamma}\xi(t).$$

After a long time, the solution converges to a limit cycle or two limit cycles depending on the period T of the amplitude of the driving field. For a short period T of the amplitude, there are two separate limit cycles, while for a long period there is only one limit cycle. In the former case, the motion is confined in each of bistable branch, and thus two separate limit cycles appear. In contrast, in the latter case, the population travels between the branches during the long period T and a unique limit cycle appears. At a threshold period, the cycles combine together and form only one limit cycle. In Fig. 6.39, we show the limit cycles near the threshold period. At $T = 0.12/\gamma$, two limit cycles appear, while at $T = 0.14/\gamma$ only one appears.

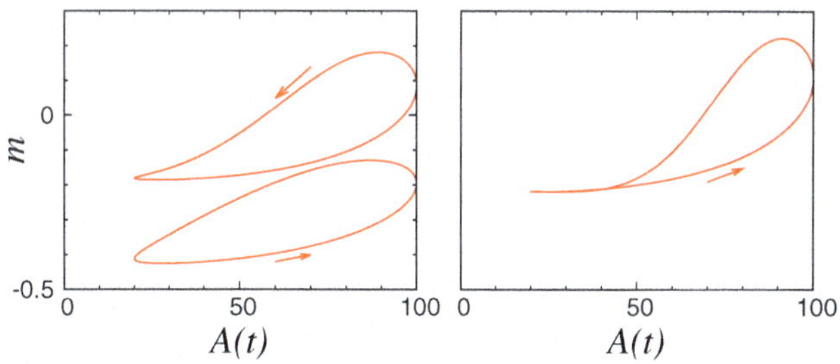

Fig. 6.39 Trajectories of limit cycles in the (A, m_z) plane obtained by the mean-field analysis (6.238) for $T = 0.12/\gamma$ and $T = 0.14/\gamma$ [143]

Part IV
Quantitative Estimation of Relaxation Time

Chapter 7
Coercivity of Magnets

7.1 Introduction

We have studied relaxation of metastable states in various cases. In this chapter, we introduce attempts to study the coercivity of magnets, which is one of the most familiar phenomena of metastability and also an important problem for practical needs in industry. Magnets are used in various places, e.g., in memory devices such as magnetic tape and hard disk, and also motors and electric generators. In particular, robots' industry rapidly grows, and joint parts of robots are made of magnets, and the demand for good magnets increases.

The strongest magnet at present is the Nd–Fe–B magnet which was invented by Dr. Masato Sagawa [146]. And currently, extensive studies for new magnets possessing further performance are also going on [147]. The strength of magnets is expressed by the product of the magnetic moment and the magnetic field. Coercivity is the amplitude of the magnetic field until which the magnetization persists in an antiparallel magnetic field. Coercivity may be called coercive field or coercive force. In principle, coercivity corresponds to the end of metastability, that is, the spinodal point (precisely speaking, the dynamical spinodal point discussed in Sect. 3.6) of materials, but it is affected by the dipole–dipole interaction and also by the fact that magnets consist of an ensemble of grains of magnetic materials.

Here, we review theoretical studies on magnetic coercivity at finite temperatures [148]. To study the coercivity of magnets quantitatively, we need to investigate the dynamics of the magnetization. Various important concepts, e.g., activation volume, magnetic after effect, and magnetic viscosity, have been introduced [35, 149]. To take into account the effects of thermal fluctuation, we study a system described by an atomistic Hamiltonian.

Here, we review two methods to estimate the coercivity at finite temperatures: a method estimating the free energy as a function of the magnetization by a Monte Carlo simulation and a method performing real-time dynamics of the magnetization using a LLG equation at finite temperature. Moreover, as a specific problem in ferromagnets,

S. Miyashita, *Collapse of Metastability*, Fundamental Theories of Physics 211,
https://doi.org/10.1007/978-981-19-6668-2_7

we refer to the effects of the dipole–dipole interaction which causes a multidomain structure in large systems and discuss the coercivity in such large grains.

As well-known, a large single crystal of magnetic material does not show coercivity even if the material has a large anisotropy. Thus, the fact that 'magnets' consist of an ensemble of small grains of the main phase and spacer materials between them (grain boundaries) is essential for the coercivity of magnets. Attempts to study such aspects will be also introduced.

7.2 Coercivity Estimated by the Free Energy Landscape

Magnetic interactions are of short range and a nucleation occurs with a finite energy gap, and, in principle, the relaxation time of the metastable state is always finite. Thus, we have to study the practical coercivity discussed in Sect. 3.6, i.e., how the relaxation time changes with the field. For example, coercivity was defined as the field at which the relaxation time of the reversal process is 1 s [150].

The coercivity of a single grain has been studied by several methods. One may obtain the free energy gap ΔF at a temperature and estimate the relaxation time by the Arrhenius relation. The rate of reversal r per 1s is given by (3.72)

$$r = N_{\text{contact}} e^{-\beta \Delta F}, \tag{7.1}$$

where N_{contact} is the frequency of contacts with the thermal bath per 1s. Using the number of contact (3.73), i.e.,

$$N_{\text{contact}} = 10^{11}, \tag{7.2}$$

we estimate the free energy gap, for which the relaxation time is 1s, as

$$1\text{s} = \frac{1}{r} = 10^{-11} e^{\beta \Delta F} \to \Delta F \simeq k_{\text{B}} T \times 25. \tag{7.3}$$

Thus, the coercivity of the 1s relaxation time is defined as the field at which the energy gap is $25 k_{\text{B}} T$.

7.2.1 Minimum Energy Path (MEP) Method

To determine the free energy gap, the so-called Minimum Energy Path (MEP) method has been used. Originally, the MEP method was introduced to obtain energy gap for a chemical reaction process [151]. In the method, a sequence of configurations is generated between a metastable state and the stable state as a kind of the path-integral method. Then, the energies along the sequence of configurations are obtained, from which we obtain the energy gap (not the free energy). This method has been extended to obtain the free energy gap for magnetization reversal using the so-called free energy functional that is an energy of spin configuration with temperature-dependent parameters [152]. This method captures characteristics of the magnetization reversal process, and it roughly gives temperature dependence of the process. However, the effect of thermal fluctuation is not taken into account precisely.

7.2.2 Free Energy Landscape Method

As an alternative method, a free energy landscape method was invented [153]. The free energy at a temperature T and a field H is given by

$$F(T, H) = -k_{\mathrm{B}}T \ln Z(T, H), \quad Z(T, H) = \mathrm{Tr} e^{-\beta\mathcal{H}_0+\beta\mu_0 H \sum_i^N S_i^z}, \tag{7.4}$$

where $\mathcal{H}_0$ is the Hamiltonian without the magnetic field. In this section, we write μ_0 explicitly. Here, we assume that the field is in the z direction which is the direction of the easy axis of the magnet,[1] and the spin along this direction at the ith site is S_i^z, and thus the term $H\sum_i^N S_i^z$ is the Zeeman energy. The total magnetization is given by

$$M = \sum_i^N S_i^z. \tag{7.5}$$

For a fixed value of the magnetization M, the free energy is given by

$$F(M : T, H) = -k_{\mathrm{B}}T \ln Z(M : T, H), \tag{7.6}$$

where

$$Z(M : T, H) = \mathrm{Tr} e^{-\beta\mathcal{H}_0+\beta\mu_0 H \sum_i^N S_i^z} \times \delta\left(\sum_i^N S_i^z - M\right). \tag{7.7}$$

This is the partition function in the space of M. The probability that the system has the magnetization M is given by

$$P(M) = \frac{Z(M : T, H)}{Z(T, H)}. \tag{7.8}$$

Thus, we can obtain $F(M : T, H)$ from the distribution function $P(m)$ as

$$F(M : T, H) - F(0 : T, H) = -k_{\mathrm{B}}T \ln \frac{P(M)}{P(0)}. \tag{7.9}$$

[1] We may apply the magnetic field not along the easy axis. But here we study the case of the field parallel to the easy axis.

A Monte Carlo simulation produces configurations proportionally to the canonical weight. Thus, a histogram of m obtained by the Monte Carlo simulation can be regarded as $P(M)$. However, the so-called important sampling Monte Carlo method generates configurations only near the equilibrium state. Because the energy of the metastable state is higher than that of the equilibrium state, $P(M)$ near the metastable state is extremely small compared to that near the equilibrium state. In a system with N spins, the ratio of probabilities has the following size dependence:

$$\frac{P(M_{\text{metastable}})}{P(M_{\text{stable}})} \sim e^{-aN}, \tag{7.10}$$

where a is a constant of order $O(1)$. Therefore, a conventional MC simulation practically does not produce states near the metastable state. Thus, it is hardly possible to obtain $P(m)$ for the entire range of m in large systems. To study the metastable state, we need precise $P(M)$ around the metastable state. To manage this difficulty, one may adopt the idea to use the so-called histogram [154, 155] to obtain states in the entire range of M by introducing an algorithm to produce spin configurations by making use of a kind of counter potential. Wang and Landau [155] proposed a practical method (Wang and Landau method) with multiple-range random walk [155].

The idea of this method is the following. First, we obtain the distribution of magnetization $P_0(M)$ by a simulation of the original Hamiltonian $\mathcal{H}_0$. The distribution is good around the equilibrium state, but not for states far from equilibrium. Then, we introduce a counter potential

$$V_0(M) = -k_{\text{B}}T \ln P_0(M), \tag{7.11}$$

and simulate the system of the modified Hamiltonian:

$$\mathcal{H}_1 = \mathcal{H}_0 + V_0(M). \tag{7.12}$$

In the system of $\mathcal{H}_1$, the states which appeared in the simulation of $\mathcal{H}_0$ are biased, and thus, other states are well sampled. Repeating this procedure, finally the distribution of the magnetization becomes flat, and the counter potential gives the free energy as a function of magnetization.

With this method, we obtained $F(M : T, 0)$ for $H = 0$ in the whole range. $F(M : T, H)$ is simply given by

$$F(M : T, H) = F(M : T, 0) - \mu_0 HM. \tag{7.13}$$

In Fig. 7.1(left), $F(M : T, H)$ obtained for the $Nd_2Fe_{14}B$ magnet at room temperature is depicted for several fields [153]. The free energy barrier F_{B} is defined as depicted in Fig. 7.1 (right). Using thus obtained free energy gap $F(M : T, H)$, one obtains the coercivity at a given temperature.

The field dependences of the free energy barrier $F_{\text{B}}(H)$ are given in Fig. 7.2. In the figure, $F_{\text{B}}(H)$ for several sizes are shown, from which we find that the size

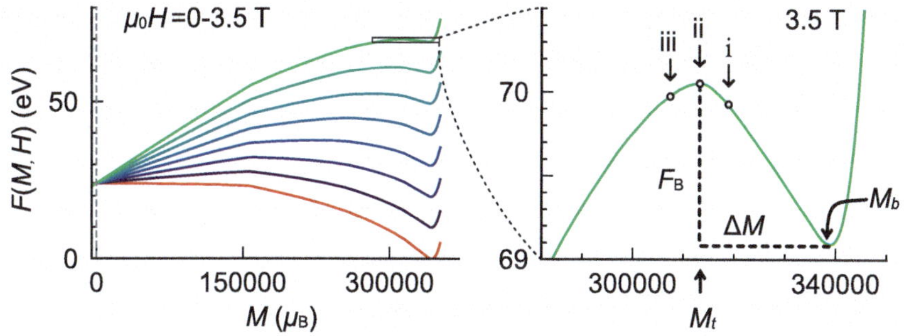

Fig. 7.1 (Left) Free energies as a function of M of the $Nd_2Fe_{14}B$ isolated grain whose size is $(L_x, L_y, L_z) = (14.1, 14.1, 14.6)$ nm (212,536 spins) at $T = 0.46\ T_C^{cal}$. $F(T, H = 0; M)$ is the bottom curve, and other lines give $F(T, H; M)$ obtained by the relation (7.13). (Right) Magnified $F(T, H; M)$ around the metastable state. Reprinted from [153] under the Creative Commons Attribution 4.0 International License, Copyright ©2020, The Author(s)

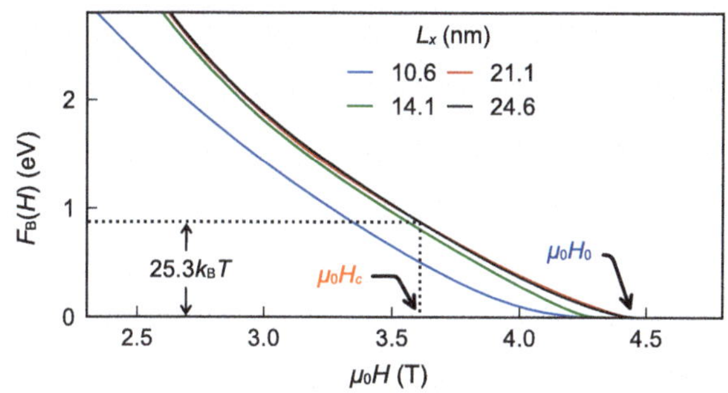

Fig. 7.2 Free energy barriers $F_B(H)$ as a function of $\mu_0 H_z$ for four system sizes: $L_x = 10.6$ nm, 14.1 nm, 21.1 nm, and 24.6 nm ($L_y = L_x$, $L_z = 1.038L_x$) at $T = 0.46T_C$. Reprinted from [153] under the Creative Commons Attribution 4.0 International License, Copyright ©2020, The Author(s)

dependence of $F(M : T, H)$ saturates for $L > 20$nm, and they are large enough to study the cases of larger sizes. F_B disappears at a certain field which we call H_0. At this field, the metastable magnetization becomes unstable. However, at finite temperatures, even before H_0 ($|H| < |H_0|$), the metastable state relaxes due to thermal fluctuation. Using the rate of relaxation rate (7.1), the free energy gap for the relaxation time 1s, the relaxation time τ is estimated by the relation (7.1) as

$$\tau = \frac{1}{r} = \frac{e^{\beta F_B}}{N_{\text{contact}}}. \tag{7.14}$$

As mentioned above, taking $N_{\text{contact}} = 10^{11}$, the free energy barrier for which the relaxation time is 1s is obtained as (7.3)

$$F_B = k_B T \times \ln(10^{11}) = 25.3 k_B T.$$

The field that gives this value of free energy barrier is called 'thermally activated coercivity', and we denote it by H_c. The value of H_c is regarded as the coercivity at the temperature.

7.3 Characteristic Quantities of Magnetization Reversal

To understand the initiation of magnetization reversal, the critical size of the nucleation of the anti-magnetized cluster plays an important role. Here, we refer to the terminologies 'activation volume' and 'magnetic viscosity' [156, 157].

7.3.1 Activation Volume

The activation volume is a volume for the critical nucleation, and it has been defined as

$$V_a = -\frac{1}{\mu_0 M_s}\frac{\partial F_B}{\partial H}, \tag{7.15}$$

where M_s is the magnetization in equilibrium. Let us consider how this quantity corresponds to the critical nucleation [153].

By differentiating the free energy barrier

$$\begin{aligned} F_B(H) =& F(M_t, H) - F(M_b, H) \\ =& F(M_t, 0) - F(M_b, 0) + \mu_0 H(M_t - M_b), \end{aligned} \tag{7.16}$$

we have

$$\frac{\partial F_B}{\partial H} = \frac{\partial F(M_t, 0)}{\partial H} - \frac{\partial F(M_b, 0)}{\partial H} + \mu_0(M_t - M_b). \tag{7.17}$$

Although $F(M_{t(b)}, 0)$ depends on H since $M_{t(b)}$ is a function of H, the contribution from these terms is zero, because $dF/dM = 0$ at $M = M_t$ and M_b, i.e.,

$$\frac{\partial F(M, 0)}{\partial H} = 0, \quad \text{at} \quad M_t \quad \text{and} \quad M_b. \tag{7.18}$$

Thus, we have

$$\frac{\partial F_B}{\partial H} = \mu_0(M_t - M_b) = \mu_0 \Delta M, \tag{7.19}$$

where ΔM is the magnetization difference for magnetic reversal, and we may call it 'activation magnetization'. Activation volume is the corresponding volume to give this magnetization difference. (7.15) gives the relation

$$V_a = \Delta M / M_s. \tag{7.20}$$

Namely, the activation volume is a volume which has the magnetization deviation ΔM from the metastable point to the top of the free energy barrier, which agrees with an intuitive picture. Here, we adopt the relation (7.20), but depending on the magnetization profile of the domain wall, the coefficient may change.

It should be noted that when the activation volume becomes large, the coercivity at finite temperature becomes small, because the effect of fluctuation of the magnetic field δH as the thermal fluctuation causes large fluctuation of Zeeman energy $\delta H \Delta M$.

7.3.2 Magnetic Viscosity

The reversal time τ of a metastable grain at a temperature T is given by (3.72) and (7.1)

$$r = \frac{1}{\tau} = Ce^{-\beta E}. \tag{7.21}$$

Here, we simply write E for the free energy barrier F_B defined in the previous subsection, and C is the constant N_{contact} in (7.1). Let us study the magnetization dynamics when the distribution of activation energy is given by $f(E)$ [149]. Initially, the number of non-flipped grains which have the activation energy $E \sim E + dE$ is

$$N(E, 0) = f(E)dE. \tag{7.22}$$

With the reversal rate (7.21), it changes as

$$\frac{dN(E)}{dt} = -CN(E)e^{-\beta E}. \tag{7.23}$$

Thus,

$$N(E, t) = N(E, 0)e^{-\lambda t}, \quad \lambda = Ce^{-\beta E}. \tag{7.24}$$

The change of magnetization by a flip of a grain being $2M_s$, the change of magnetization is given by the sum of contributions from all E.

$$\frac{dM}{dt} = 2M_s C \int_{E_{\min}}^{E_{\max}} f(E, 0)e^{-\lambda t}e^{-\beta E} dE. \tag{7.25}$$

Here, $E_{\min}$ and $E_{\max}$ give the range of energy gap distribution, and here we adopt $E_{\min} = 0$ and $E_{\max} = \infty$. When we assume that $f(E)$ is almost constant around the relevant energy range for a given time scale τ, putting $f(E) \simeq p$, we may approximate (7.25) as

$$\frac{dM}{dt} = 2M_s Cp \int_0^{\infty} e^{-\lambda t} e^{-\beta E} dE = -2M_s Cp \int_{\lambda=C}^{\lambda=0} e^{-\lambda t} \frac{d\lambda}{\beta C} = 2M_s k_B T p \left[\frac{1}{t} e^{-\lambda t} \right]_{\lambda=C}^{\lambda=0}, \tag{7.26}$$

where we used the relation $\lambda = Ce^{-\beta E}$. For large Ct, the magnetization change is given by

$$M(t) - M(0) = \int_0^t \frac{dM}{dt} dt = 2M_s p k_B T \ln t + \text{const.} = S \ln t + \text{const.} \tag{7.27}$$

This logarithmic change of magnetization is due to the distribution of activation energy, and the coefficient S is called 'magnetic viscosity' [149].

$$S = 2M_s k_B T f(E_\tau) \tag{7.28}$$

where E_τ is the typical energy barrier to give a given relaxation time τ, which is given by (7.21) for $C = 10^{11}$ and $\tau = 1$s:

$$\frac{1}{\tau} = Ce^{-\beta E_\tau} \rightarrow E_\tau = k_B T \ln C\tau \simeq 25 k_B T. \tag{7.29}$$

Phenomena of magnetic viscosity have been observed in various systems of permanent magnets, spin glass systems [158], and systems consisting of magnetic nanoparticles [159].

7.3.3 Relation Between the Activation Volume V_a and the Magnetic Viscosity S

The magnetic viscosity coefficient can be related to the activation volume [156, 160]. Let the distribution of energy barrier for a given field H be $f_H(E)$. For the relaxation time τ, the part $E < E_\tau$ is flipped and the magnetization is given by

$$M = M_s \left(2 \int_{E_\tau}^{\infty} f(E) dE - 1 \right). \tag{7.30}$$

If $\tau = 0$, $E_\tau = -\infty$ then all the grains do not flip and $M = M_s$, while $\tau = \infty$, $E_\tau = \infty$ then all the grains flip and $M = -M_s$. When the field is increased $H \rightarrow H + \Delta H$, the distribution changes to a new distribution $f_1(E)$ and the magnetization is given by

$$M = M_s \left(2 \int_{E_\tau - \Delta E}^{\infty} f_1(E) dE - 1 \right), \tag{7.31}$$

where

$$\Delta E = -\left(\frac{dE}{dH} \right) \Delta H. \tag{7.32}$$

The change of magnetization after waiting time τ is

$$\Delta M = 2M_s \int_{E_\tau - \Delta E}^{E_0} f_1(E) dE = -2M_s f_1(E) \left(\frac{dE}{dH} \right) \Delta H. \tag{7.33}$$

The irreversible susceptibility for waiting time τ is defined as

$$\chi_{\rm irr} = \left(\frac{dM}{dH} \right)_\tau = -2M_s f_1(E) \left(\frac{dE}{dH} \right). \tag{7.34}$$

This quantity can be obtained by the remanent magnetization loop. Comparing with (7.28), we have the relation:

$$S = -k_{\rm B} T \chi_{\rm irr} \left(\frac{dE}{dH} \right)^{-1}. \tag{7.35}$$

Using the definition of the activation volume (7.15), we have

$$S = \frac{k_{\rm B} T \chi_{\rm irr}}{\mu_0 M_s V_{\rm a}}. \tag{7.36}$$

The quantity defined by

$$S_v \equiv \frac{S}{\chi_{\rm irr}} \tag{7.37}$$

is often used [160], which is given by

$$S_v = -k_{\rm B} T \left(\frac{dE}{dH} \right)^{-1} = \frac{k_{\rm B} T}{\mu_0 M_s V_{\rm a}}. \tag{7.38}$$

This quantity is sometimes called the fluctuation field:($S \equiv \chi_{\rm irr} H_{\rm f}$)

$$H_{\rm f} = S_v. \tag{7.39}$$

7.3.4 Coercivity Obtained by a Direct Simulation of SLLG

One may approach the coercivity more directly by a numerical simulation using the so-called the stochastic Landau–Lifshitz–Gilbert (SLLG) equation. To take into account the thermal fluctuation [161, 162], the SLLG equation includes a white-Gaussian random field $\{\boldsymbol{\xi}_i\}$:

$$\frac{d}{dt}\boldsymbol{S}_i = -\frac{\gamma}{1+\alpha_i^2}\boldsymbol{S}_i \times (\boldsymbol{H}_i^{\text{eff}} + \boldsymbol{\xi}_i) - \frac{\alpha_i\gamma}{(1+\alpha_i^2)S_i}\boldsymbol{S}_i \times (\boldsymbol{S}_i \times (\boldsymbol{H}_i^{\text{eff}} + \boldsymbol{\xi}_i)), \tag{7.40}$$

where $\gamma = g\mu_{\text{B}}/\hbar$ is the gyromagnetic constant, and α_i is the damping factor at the ith site. The effective field due to the exchange interaction and the anisotropy terms is given by

$$\boldsymbol{H}_i^{\text{eff}} = -\frac{\partial \mathcal{H}}{\partial \boldsymbol{S}_i}, \tag{7.41}$$

where $\mathcal{H}$ is the Hamiltonian of the system. For the random field representing the thermal noise, we use the white-Gaussian noise $\boldsymbol{\xi}_i$ which satisfies

$$\langle \xi_k^{\mu}(t)\rangle = 0, \quad \langle \xi_k^{\mu}(t)\xi_l^{\nu}(s)\rangle = 2\tilde{D}_k\delta_{kl}\delta_{\mu\nu}\delta(t-s), \quad \mu = x, y, \text{ or } z. \tag{7.42}$$

The strength of the noise, $\tilde{D}_k$, is related to the temperature of the system by the fluctuation–dissipation relation

$$\tilde{D}_k = \frac{\alpha_k}{S_k}\frac{k_{\text{B}}T}{\gamma}, \tag{7.43}$$

where S_k is the amplitude of the magnetization of kth spin. For the damping constant α_i, we adopted a commonly accepted value. The dependence of the relaxation time on the damping constant is discussed later.

By using the SLLG equation (7.40), the dynamics of magnetization of $Nd_2Fe_{14}B$ was studied [163, 164]. Snapshots of configurations during a reversing process from a uniform ferromagnetic state with all down-spins in an opposite field $H = 4.0$ T are depicted in Fig. 7.3a. In the figure, we find a nucleation occurring from a corner.

In Fig. 7.3b, we depict some samples of dynamics of the magnetization in relaxation processes at a high field $H = 8$T (Fig. 7.3b (left)) and at a low field $H = 4.1$T (Fig. 7.3b (right)). Respectively, they represent the deterministic and the stochastic types of relaxations, which are explained in Sect. 3.6, and schematic pictures are depicted in Fig. 7.4.

In the deterministic type (Fig. 7.3b (left)), the width of the distribution of the relaxation times is small. On the other hand, in the stochastic type (Fig. 7.3b (right)), the relaxation is triggered by single nucleation, and the width of the distribution of the relaxation times is very wide. In Fig. 7.3b (right), we find a few samples which

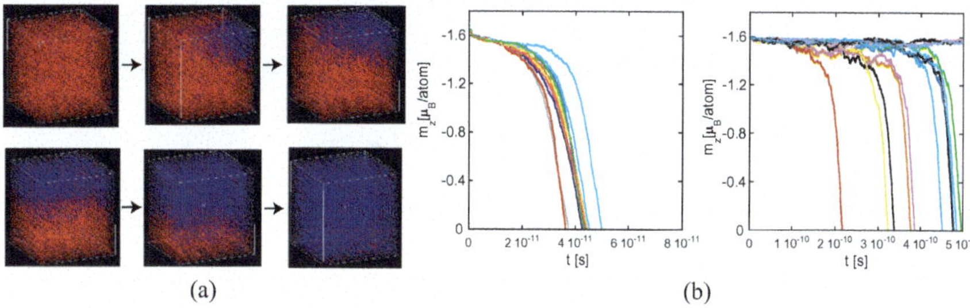

Fig. 7.3 **a** Snapshots during a magnetization reversal from the all-down- spin state under a reversed field ($H = 4{:}0$ [T]). **b** Examples of time evolutions of magnetization relaxation curves at (left) $H = 8$T and (right) $H = 4.1$T. $\alpha = 0.1$. From reference [163]: modified

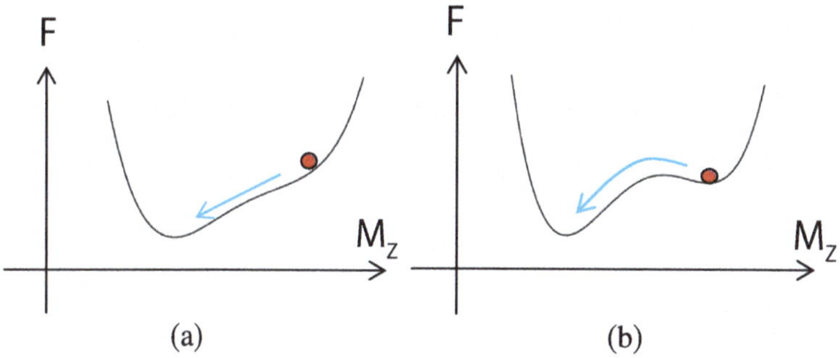

Fig. 7.4 **a** Deterministic process. **b** Stochastic process by barrier-crossing dynamics

do not relax yet. To obtain the average relaxation time, we must wait until all the samples relax. Thus, this wide distribution makes it difficult to estimate the average relaxation time.

To overcome this difficulty in the determination of the average relaxation time, we make use of a statistical method for the distribution of the relaxation times. Let the probability of a nucleation in unit time be p. Then, the probability $P(t)\Delta t$ that the first event occurs in the period $[t; t + \Delta t]$ is given by

$$P(t)\Delta t = pe^{-pt}\Delta t, \tag{7.44}$$

and the mean relaxation time τ is given by

$$\tau = \langle t \rangle = p \int_0^\infty te^{-pt}dt = \frac{1}{p}. \tag{7.45}$$

When we perform simulations of many samples, the number of surviving (unchanged) samples is

$$N_{\rm sv}(t) = N - N_{\rm done}(t) = Ne^{-pt}, \tag{7.46}$$

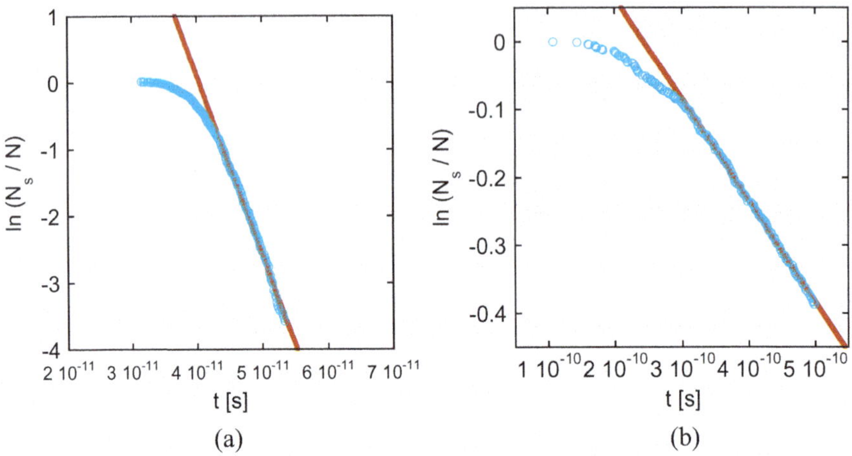

Fig. 7.5 Time dependence of $\ln(N_s/N)$. Circles denote time dependence of $\ln(N_s/N)$. **a** H = 8T and (b) H = 4.1T. Here the damping constant is $\alpha = 0.1$. From reference [163]: modified

where N is the number of samples. From this relation, p, i.e., τ, can be estimated from the slope of $\ln(N_{sv}(t)/N)$, which is depicted in Fig. 7.5. In the figure, we recognize a time range of linear dependence in which the dynamics is governed by a single nucleation process (7.44). In the early stage, the data deviate from the line, which indicates some early-time effect on the nucleation, which should be removed when we estimate the value of p. This recognition of deviation is difficult in the estimation of relaxation time by a naive average over the samples.

Another difficulty in the estimation of relaxation time is a large dependence of the relaxation time on the damping constant α. We may regard α as a time scale and may expect the relaxation time is proportional to $1/\alpha$. However, for the process of nucleation, the situation is not so simple. In Fig. 7.6a, we show the field dependence of relaxation time for different values of α, and find a large dependence. Because we do not know a precise value of α, it is very difficult to estimate the relaxation time quantitatively.

In this circumstance, we may use the fact that the relaxation time increases very rapidly below a certain field, say $H \simeq 4.2$ T. There, although the distribution of the relaxation times at a given field is wide, the distribution of time for a given value of the relaxation time is small because of the large slope. Therefore, the value of the field for a given relaxation time can be estimated with small errors. We expect that the sharp growth of relaxation time is governed by dynamics jumping over the free energy barrier which may be a simple exponential function, and we fit them by the following form of an exponential decay with a correction, i.e., a double exponential form:

$$\tau(h) = Ae^{-aH} + Be^{-bH} = Ae^{-aH}\left(1 + Ce^{-dH}\right), \tag{7.47}$$

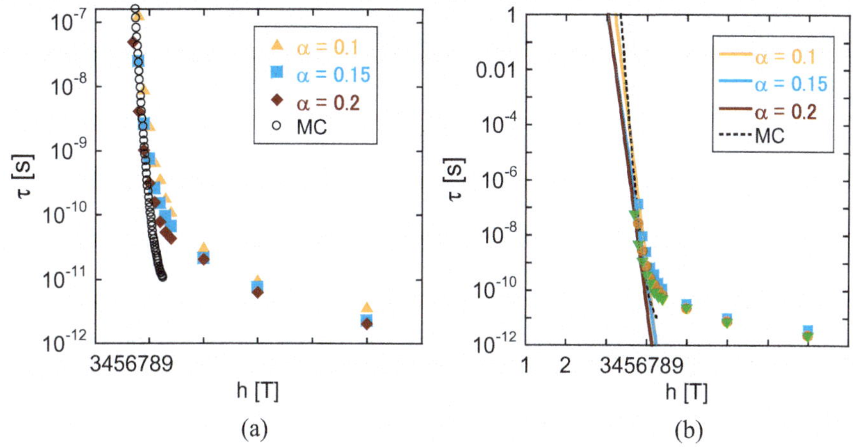

Fig. 7.6 **a** Magnetic field dependence of the relaxation time for several values of the damping factor α. Open circles give field dependence of the relaxation time assuming an Arrhenius law ($\tau = \tau_0 e^{\Delta F}$) with the value of ΔF taken from the Monte Carlo study for $L_x = 10.6$ nm in Fig. 7.2. **b** Extrapolation of the relaxation time to estimate the coercivity with 1s persisting time. From reference [163]: modified

where $C = B/A$ and $d = b - a$. In Fig. 7.6b, the fitted curves by Eq. (7.47) are plotted for several values of α. The field at which $\tau = 1$s gives the coercivity with 1s persisting time. The estimated coercivities for $\alpha = 0.1, 0.15$, and 0.2 are $H_c \simeq 3.2$, 3.0, and 3.0, respectively. From these, we estimate the coercivity as

$$h_c \simeq 3\text{T}. \tag{7.48}$$

This value is close to the estimation $h_c \simeq 3.3$ obtained by the MC method in the same size [153] explained in the previous section.

7.3.5 *Coercivity of Large Grains*

In previous subsections, we studied coercivity without the dipole–dipole interaction (DDI).

$$V_{\text{DDI}}(\boldsymbol{s}_i, \boldsymbol{s}_j) = \mathcal{D}\left(\frac{\boldsymbol{s}_i \cdot \boldsymbol{s}_j}{r_{ij}^3} - 3\frac{(\boldsymbol{s}_i \cdot \boldsymbol{r}_{ij})(\boldsymbol{s}_j \cdot \boldsymbol{r}_{ij})}{r_{ij}^5}\right). \tag{7.49}$$

Ferromagnetic systems have a uniform magnetization, in which DDI causes a relevant effect on the ordered configuration when the size of a system increases. For example, DDI causes the demagnetization field, and also deforms the uniformly ordered state to multidomain states [165, 166]. When the system size increases, DDI becomes an important factor to determine the coercivity. We have studied the effects of DDI in thin films of the present material $Nd_2Fe_{16}B$ [167].

7.3.5.1 Stochastic Cutoff (SCO) Method

Because DDI is an interaction among all the spins, the computational cost to calculate the energy is proportional to the square of the number of spins (N^2). To overcome this difficulty, various methods have been proposed [168]. The so-called stochastic cutoff (SCO) method has also been proposed as one of them [169, 170]. The SCO method introduces a selection of bonds by the so-called switching procedure. The selection procedure is performed stochastically, maintaining the detailed balance condition. Thus, the stationary state of the simulation is guaranteed to be the same as the equilibrium state of the original model [169, 171]. Because the bond selection process rarely adopts long-distant weak bonds, the number of selected bonds is small and the overall computational time is markedly decreased. However, for the study of $Nd_2Fe_{14}B$ which has a unit cell consisting of 68 atoms, most of the previous methods are not very efficient because of the large number of spins in a unit cell. To manage this difficulty in complicated systems, a modified SCO method by using Walker's algorithm[170] has been introduced [172].

7.3.5.2 Multidomain State Due to DDI

Because of DDI, for large systems, a uniform ferromagnetic configuration is no longer stable and a multidomain magnetic structure is formed. To study the general tendency of dependence of the magnetic patterns on the anisotropy coefficient K and the strength of the DDI D, the following simplified model was studied [148, 167]:

$$\mathcal{H} = -\sum_{i,j} J_{ij} \boldsymbol{s}_i \cdot \boldsymbol{s}_j - \sum_i K s_{zi}^2 + \sum_{i,j} V_{\mathrm{DDI}}(\boldsymbol{s}_i, \boldsymbol{s}_j), \tag{7.50}$$

where $\boldsymbol{s}_i$ is a unit vector.

In Fig. 7.7, magnetic profiles of systems of $64 \times 64 \times 10$ are depicted in a phase diagram in the space $(K/J, \mathcal{D}/J)$. The left figure is obtained by a thermal demagnetization process, i.e., a process in which the temperature is quenched to $0.3T_C$ starting from a random configuration. On the other hand, the right figure is obtained by a field-quench process, i.e., a process in which the field is quenched to zero starting from a large value of the field which brings the system to a complete ferromagnetic state.

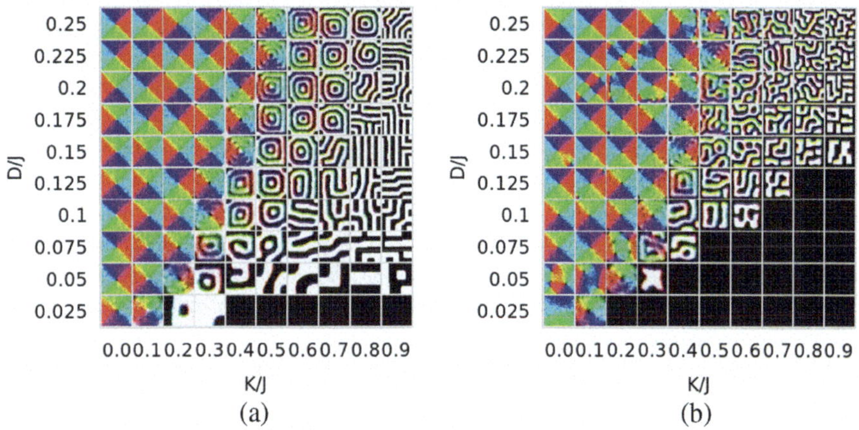

Fig. 7.7 **a** Magnetic structures for $(K/J, \mathcal{D}/J)$ obtained by the thermal-quench process at $T = 0.3T_C$. (T_C is the critical temperature of the bulk system.) **b** Magnetic structures obtained by the field-quench process. Spin direction (s_x, s_y, s_z) is coded in the manner shown in the right in Fig. 7.8. Reprinted from [148] under the Creative Commons Attribution 4.0 International License, Copyright ©2021, The Author(s)

7.3.5.3 Coercivity of Grains of Multidomain Size

In both figures, we find five typical magnetic patterns, i.e., out-of-plane (along the z axis) ferromagnet, in-plane ferromagnet, vortex, multidomain, and canted multidomain. In the right figure, the region of the out-of-plane ferromagnetic region is larger than that in the left figure, which indicates that the ferromagnetic state can remain in the field-quench process at some parameter $(K, \mathcal{D})$ sets in which the multidomain appears in the thermal-quench process. This implies that the ferromagnetic state, i.e., the state with aligned spins, persists as a metastable state in the region.

To study this metastability as a function of $\mathcal{D}$, we study how the ferromagnetic state survives when $\mathcal{D}$ increases starting from the ferromagnetic region, e.g., a point $K = 0.7$ and $\mathcal{D}/J = 0$. When $\mathcal{D}$ reaches a threshold value, the nucleation of the maze occurs. The configurations immediately after the nucleation are depicted in Fig. 7.8. It should be noted that the nucleation in the present case begins inside of the plane, not from corners, which was found in a nanoscale system as shown in Fig. 7.3 (left).

Here, we find the ferromagnetic state persists as a metastable state although the true stable state is a multidomain state. There, the hysteresis between the ferromagnetic state and the multidomain state exists [165, 166], in addition to the hysteresis between ferromagnetic states of positive and negative magnetizations.

This kind of multistep phase transition can be studied by a simplified mean-field analysis assuming two order parameters x and y which represent order parameters with positive and negative magnetizations. The short-range interaction is ferromagnetic which tends to align spins parallel. When magnetic clusters are set in the

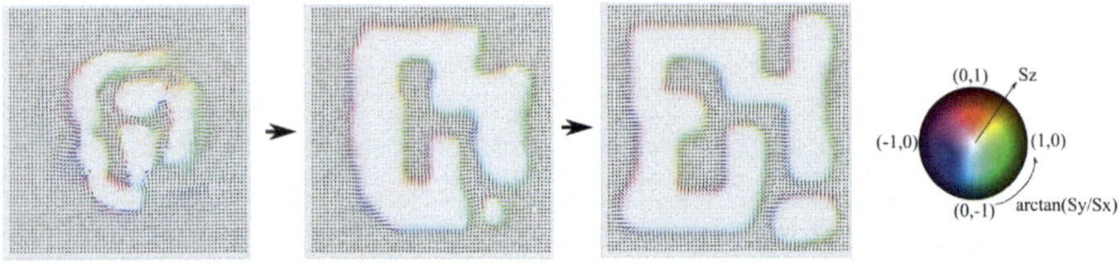

Fig. 7.8 Nucleation patterns just after the nucleation. The spin direction (s_x, s_y, s_z) is coded in the manner shown in the right. Reprinted from [148] under the Creative Commons Attribution 4.0 International License, Copyright ©2021, The Author(s)

perpendicular direction to the magnetization, DDI tends to make the cluster antiparallel. On the other hand, when magnetic clusters are set in the parallel direction to the magnetization, DDI tends to make the cluster parallel. Here, the magnetization is perpendicular to the plane, and thus the effect of DDI is taken into account as an interaction to make the magnetizations of cluster antiparallel. Taking into account these interactions, we adopt the following simplified Hamiltonian:

$$\mathcal{H} = -J \sum_{<i,j>\text{cluster A}} S_i \cdot S_j - J \sum_{<k,l>\text{cluster B}} S_k \cdot S_l + \mathcal{D} \sum_{i,k} S_i^{\mathrm{A}} \cdot S_k^{\mathrm{B}} - H \sum_i S_i^z - H \sum_k S_k^z, \tag{7.51}$$

where J denotes the short-range ferromagnetic interaction, $\mathcal{D}$ represents the effect of DDI, and H is the external field.

Setting $x = \langle S_i^z \rangle$ and $y = \langle S_k^z \rangle$, the mean-field free energy of the model is given by

$$F(x, y) = \frac{Jz}{2}x^2 - k_{\mathrm{B}}T \ln\left[2\cosh(\beta Jzx - \beta\mathcal{D}y + \beta H)\right]$$

$$+ \frac{Jz}{2}y^2 - k_{\mathrm{B}}T \ln\left[2\cosh(\beta Jzy - \beta\mathcal{D}x + \beta H)\right] \tag{7.52}$$

as has been studied in the spin-crossover problem [24]. In Fig. 7.9, typical magnetization processes $M(H) = x + y$ are depicted. At the high temperature $T/J = 8.0$, the system is in a paramagnetic state (i.e., disordered state) and $M(H)$ is monotonic. At $T/J = 6.0$, the hysteresis between the paramagnetic phase and ferromagnetic phase appears. At the low temperature $T/J = 4.0$, the ferromagnetic phase shows coercivity.

7.4 Coercivity of Magnets as an Ensemble of Grains

So far, we studied the metastable state of a single grain of magnet. But real magnets consist of grains of magnetic material. Indeed, as we mentioned, a large single crystal of magnetic material does not show coercivity. Therefore, the metastability of real

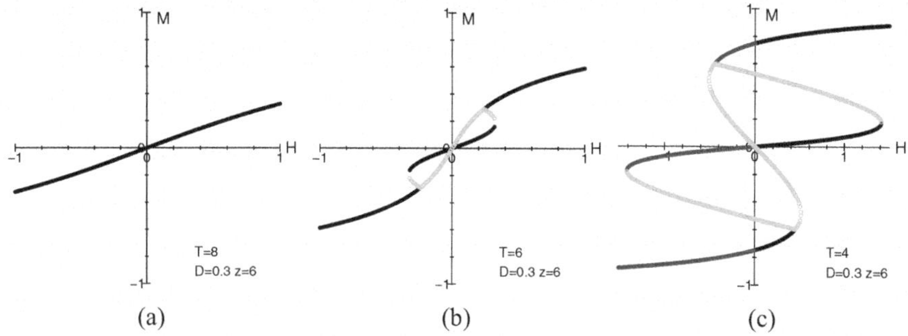

Fig. 7.9 Magnetization processes $M(H) = x + y$ (7.52) for $\mathcal{D} = 0.3$ and $z = 6$. **a** $T/J = 8.0$, **b** $T/J = 6.0$, and **c** $T/J = 4.0$. Dots show stable solutions and open circles show unstable solutions of the mean-field analysis

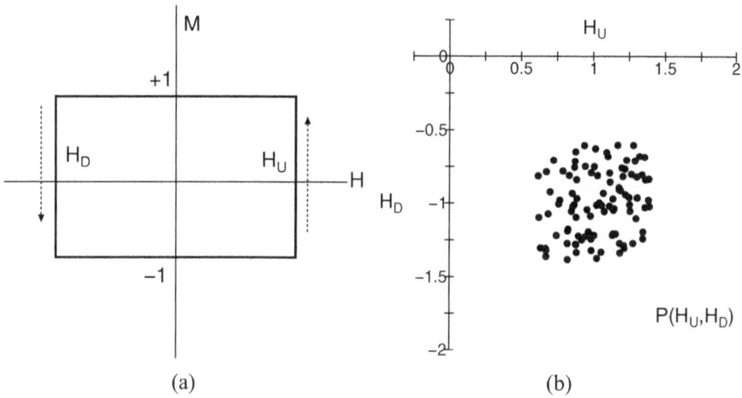

Fig. 7.10 **a** A hysteron with the thresholds H_D and H_U. **b** An example of the distribution of H_D and H_U: $P(H_D, H_U)$

magnets, namely coercivity, must be studied by taking into account the effects of the ensemble of grains.

Each grain has its own coercivity. Let us consider a simplified hysteresis which has a threshold field H_D at which magnetization changes from $+1$ to -1, and another threshold field H_U at which magnetization changes from -1 to $+1$, as depicted in Fig. 7.10a. A unit having this hysteresis is called 'hysteron'.

Let us consider an ensemble of hysterons which have the thresholds $\{(H_D(i), H_U(i))\}$, $i = 1, \ldots N$. There, H_D and H_U are distributed. Let the distribution function of (H_D, H_U) be $P(H_D, H_U)$. An example of $P(H_D, H_U)$ is depicted in Fig. 7.10b. Here, we assume that all the grains do not interact. The magnetization of the total system is given by the sum of those of grains $\sigma_i (= \pm 1)$

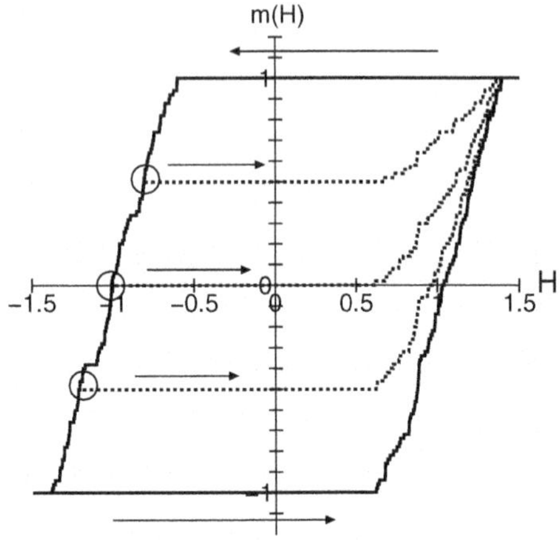

Fig. 7.11 Magnetization curve in a swept field. The bold one gives the full hysteresis for which the field H is swept from $1.5 \to -1.5 \to 1.5$. The dotted lines show the magnetization when the field turns back at the field H_r shown by circles (first-order reversal curve (FORC))

$$M(H) = \sum_{i=1}^{N} \sigma_i, \quad m(H) = \frac{M(H)}{N}. \tag{7.53}$$

Let all the hysterons be $+1$ ($\sigma_i = 1$) at a large value of H. The magnetization changes ($m = 1 \to -1$) when the field is swept from the large value to the negative value. The magnetization changes at every $H_{\rm D}(i)$. When H becomes smaller than the minimum value of $\{H_D(i)\}$, $m(H)$ reaches -1. Then, the field is swept back. When H reaches the minimum value of $\{H_U(i)\}$, say $H_U(j)$, the jth hysteron changes the spin: $\sigma_i = -1 \to +1$, and at every $H_{\rm U}(i)$ M increases, and at the maximum value of $H_U(i)$, m becomes $+1$. In this way, the magnetization shows a hysteresis according to the distribution $P(H_{\rm D}, H_{\rm U})$ as depicted in Fig. 7.11. This model is called the Preisach model [173].

If we reverse the field at H_r before all the spins flip down, $M(H)$ starts increasing. But $M(H)$ does not follow back the $M(H)$ in the process of decreasing field, and show a hysteresis. Examples of $M(H_r, H)$, $H \geq H_r$ are depicted by dotted lines in Fig. 7.11. This hysteresis is naturally understood from the fact that the ith hysteron which flipped at $H_{\rm D}(i)$ remains to be -1 until the field reaches $H_{\rm U}(i)$. This type of hysteresis as a function of hysteron H_r is called the 'first-order reversal curve (FORC)'.

From the FORCs, $\{M(H_r, H)\}$, of all the values of H_r, we can obtain the distribution function $P(H_{\rm D}, H_{\rm U})$ by the relation:

$$P(H_1 = H_r, H_2) = \left. \frac{\partial^2 M(H_r, H)}{\partial H_r \partial H} \right|_{H=H_2}. \tag{7.54}$$

The distribution $P(H_{\rm D}, H_{\rm U})$ obtained by the above relation is called the 'FORC diagram', which should be the same as $P(H_{\rm D}, H_{\rm U})$ if the hysterons do not interact.

In real magnets, grains of the hard magnet correspond to the hysteron, but they interact with each other. Thus, we cannot use the relation (7.54). However, $P(H_D, H_U)$ thus obtained still gives important information of magnets, and the FORC diagram is used to characterize the features of magnets [174–176].

How the magnetization curve changes with types of interaction among the grains is an interesting problem. The effect of DDI on FORCs has been taken into account within a mean-field-type analysis, which is called the 'moving (Preisach) model' [177]. Taking into account the distributions of alignment of the easy axis, DDI among grains and effects coming from neighboring grains through the grain boundary phase, one may survey types of magnetic curves of an ensemble of grains. In real magnets, each grain may be large, and the multi-magnetic domain structure appears in it. For these grains, the field dependence of magnetization curve $M(H)$ of the hysteron is no more given by the simple shape as depicted in Fig. 7.10 but by a smooth shape. The smooth response to the field is called the 'reversible component'. Moreover, $M(H)$ may have multi-branch as depicted in Fig. 7.9b and c. To capture detailed characteristics of magnetization processes of various real magnets, these facts also need to be taken into account [178].

Part V
Appendices

Chapter 8
Appendices

8.1 Brief Review on the Mean-Field Approximation

Because the free energy $f(m: T, H)$ as a function of order parameter m plays an important role to study the metastability, we review how to obtain $f(m: T, H)$ by the mean-field analysis for the Ising model.

8.1.1 *Basic Idea of Mean-Field Theory*

The Hamiltonian of the ferromagnetic Ising model is given by

$$\mathcal{H} = -J \sum_{<ij>} \sigma_i \sigma_j - H \sum_i \sigma_i, \quad \sigma = \pm 1. \tag{8.1}$$

Here, we assume that the spin σ_i at the site i interacts with the spins σ_j at its nearest neighbor sites. Let the number of nearest neighbor sites be z. For example, $z = 4$ in the square lattice and $z = 6$ in the simple-cubic lattice. In the mean-field approximation, we replace the spins in the neighboring sites $\{\sigma_j\}$ by the average magnetization m (Fig. 8.1). Then, the Hamiltonian for σ_i is given by

$$H_i = -Jzm\sigma_i - H\sigma_i. \tag{8.2}$$

Then, we obtain the expectation value of σ_i at a temperature T in this Hamiltonian.

$$\langle \sigma_i \rangle = \frac{\sum_{\sigma_i=\pm 1} \sigma_i e^{\beta(zJm+H)\sigma_i}}{\sum_{\sigma_i=\pm 1} e^{\beta(zJm+H)\sigma_i}} = \tanh(\beta Jzm + \beta H), \quad \beta = \frac{1}{k_B T}. \tag{8.3}$$

Requiring the obtained value of $\langle \sigma_i \rangle$ to be the same as the assumed magnetization m of the neighboring sites, we have the 'self-consistent equation', which is (2.4) in the text:

S. Miyashita, *Collapse of Metastability*, Fundamental Theories of Physics 211,
https://doi.org/10.1007/978-981-19-6668-2_8

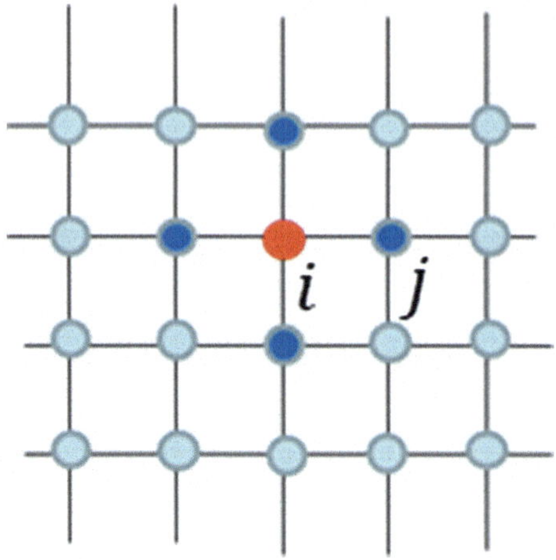

Fig. 8.1 Mean-field approximation $\sigma_j \to m$. The spins at neighboring sites $\{\sigma_j\}$ of the site i are replaced by the mean of magnetization m

$$m = \tanh(\beta Jzm + \beta H). \tag{8.4}$$

This relation can be rewritten as

$$\beta Jzm + \beta H = \tanh^{-1} m = \frac{1}{2} \ln \frac{1+m}{1-m} \to H = \frac{k_B T}{2} \ln \frac{1+m}{1-m} - Jmz, \tag{8.5}$$

which is depicted in Fig. 2.2. The critical temperature is the point where the system begins to have nonzero solutions $\pm m_s$ at $H = 0$ (2.7), which are solutions of

$$m = \tanh(\beta Jzm). \tag{8.6}$$

The nonzero solution is called 'spontaneous magnetization'. Near the temperature, the nonzero solution m is small. Expanding (8.6) with respect to m:

$$m = \beta Jzm - \frac{1}{3}(\beta Jzm)^3 \cdots \to m = 0, \quad m_s \sim \pm\sqrt{\frac{3(\beta Jz - 1)}{(\beta Jz)^3}}, \tag{8.7}$$

and the critical temperature is found to be

$$k_B T_C = zJ. \tag{8.8}$$

Below the critical point, the system has hysteresis as we see in Fig. 2.2.

Here, we studied the solution of (8.6) by the analysis of (8.5) at $H = 0$; the solutions of (8.6) are also obtained as cross points of the functions:

$$\begin{cases} y = m \\ y = \tanh(\beta(Jzm + H)) \end{cases} \tag{8.9}$$

For $H = 0$, the functions have trivial solution $m = 0$, and in addition the spontaneous magnetization $\pm m_s$ below the critical point $\beta J z > 1$.

Thus, there are three solutions below the critical point, and we need to determine which one represents the true equilibrium. Therefore, we obtain the free energy as a function of magnetization in the next subsection.

8.1.2 Mean-Field Free Energy as a Function of the Magnetization $F(m : T, H)$

The free energy F is obtained from the partition function Z as

$$F = -k_B T \ln Z. \tag{8.10}$$

But, in the mean-field analysis we need to be careful. That is, if we adopt the sum of one-body Hamiltonian (8.2) as the total energy:

$$\mathcal{H}_{\mathrm{MF}} \equiv \sum_{i=1}^{N} \mathcal{H}_{\mathrm{MF}}^{(i)} = -Jz \sum_{i=1}^{N} m\sigma_i - H \sum_{i=1}^{N} \sigma_i, \tag{8.11}$$

the free energy is obtained as

$$F_{\mathrm{MF0}}(m) = -k_B T \ln Z_{\mathrm{MF0}} = -k_B T N \ln\left(2\cosh(\beta z m + \beta H)\right). \tag{8.12}$$

This free energy has a single maximum at $m = 0$ regardless of the temperature. This strange behavior comes from the fact that the Hamiltonian (8.11) only represents local interaction and does not express the total energy of the system (8.1) correctly.

To obtain the free energy for the system (8.1) in the mean-field analysis, we need to construct appropriate approximations for the total system. Some of them are introduced below.

In the studies on the first-order phase transition, the free energy as a function of the magnetization $F(m : T, H)$ is often used as a potential energy for the order parameter. That is, dynamics of the order parameter (van Hove equation) is often given by

$$\frac{dm}{dt} = -\frac{1}{\tau_0}\frac{df(m)}{dm} + \text{noise term}. \tag{8.13}$$

The equation comes from the overdamped approximation for the equation of motion of a particle of the mass M with the friction term

$$M\frac{d^2x}{dt^2} = -k\frac{dx}{dt} - \frac{dU}{dx} = 0 \rightarrow \frac{dx}{dt} = -\frac{1}{k}\frac{dU}{dx} \tag{8.14}$$

by setting $k = \tau_0$. There, the form of $f(m)$ determines the dynamics, and we have to be careful about the form of $f(m)$. Therefore, we explore the ways to obtain $F(m:T,H)$ rather in detail.

8.1.2.1 Approximation Ignoring the Fluctuation of Magnetization

Let us express σ_i as sum of the mean m and the fluctuation δ_i:

$$\sigma_i = m + \delta_i, \quad \delta_i = \sigma_i - m. \tag{8.15}$$

Substituting this relation to the interaction part of the original Hamiltonian (2.1)

$$\begin{aligned} -J\sum_{<ij>} \sigma_i\sigma_j &= -J\sum_{<ij>}(m + (\sigma_i - m))(m + (\sigma_j - m)) \\ &= -J\sum_{<ij>}(-m^2 + m(\sigma_i + \sigma_j) + (\sigma_i - m)(\sigma_j - m)) \\ &= +\tfrac{z}{2}Jm^2 - Jmz\sum_i \sigma_i + \sum_{<ij>}(\sigma_i - m)(\sigma_j - m). \end{aligned} \tag{8.16}$$

Up to here, no approximation has been done.

Now, we ignore the second order of the fluctuation

$$(\sigma_i - m)(\sigma_j - m)) \to 0. \tag{8.17}$$

Then, we have a Hamiltonian of the mean-field approximation:

$$\hat{\mathcal{H}}_{\text{MF}} = \frac{zJNm^2}{2} - (Jzm + H)\sum_i \sigma_i. \tag{8.18}$$

The partition function is easily calculated as

$$Z = \sum_{\sigma_i=\pm 1} e^{-\beta\mathcal{H}_{\text{MF}}} = e^{-\beta\frac{zJNm^2}{2}}\left(e^{\beta(zJm+H)} + e^{-\beta(zJm+H)}\right)^N \tag{8.19}$$

and the free energy in this approximation is

$$F(m:T,H) = \frac{zJNm^2}{2} - k_{\text{B}}TN\ln(2\cosh(\beta(zJm + H))). \tag{8.20}$$

The extrema of $F(m:T,H)$ is given by

$$\left(\frac{\partial F}{\partial m}\right)_{T,H} = zJN\,(m - \tanh(\beta(zJm + H))) = 0 \tag{8.21}$$

which coincides with the self-consistent equation (8.4). Thus, this form correctly reproduces the mean-field relation and can be regarded as a free energy of a function

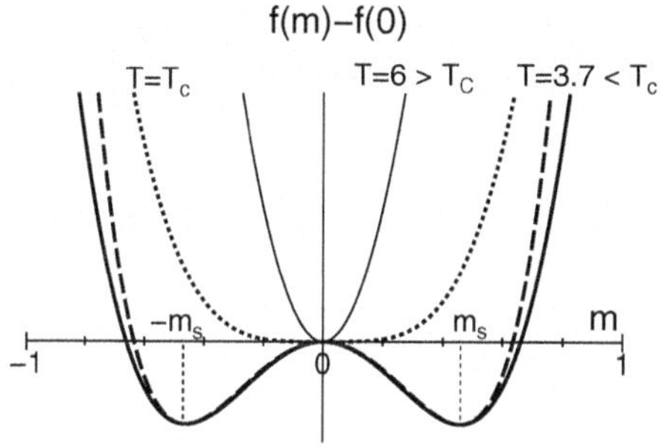

Fig. 8.2 Free energy in the mean-field approximation (8.20) is plotted as $T > T_C$(thin line), $T = T_C$(dotted line), and $T < T_C$(bold line). The dashed line plots the free energy $F_{BW}(m)$ at $k_B T = 3.7J < k_B T_C$ which will be explained in the next subsection. (In the figure, k_B is not written.)

of magnetization. In Fig. 8.2, the form is plotted for $T > T_C$(thin line), $T = T_C$(dotted line), and $T < T_C$(bold line). Here, we find the free energy for $\pm m_s$ is lower than that at $m = 0$. Thus, we know that one of the states of $\pm m_s$ gives the equilibrium state below T_C and $m = 0$ gives an unstable solution, while $m = 0$ is a unique stable solution above T_C.

Here, the following point should be noted. Although we call $F(m:T,H)$ free energy, it is not the free energy in the sense of thermodynamics $F(T,H)$ which is a function of T and H, but not of m. The value of $F(m:T,H)$ at the minimum point gives $F(T,H)$.

As we see here, $F(m:T,H)$ plays a kind of potential energy for the order parameter, and it is called phenomenological free energy or Landau free energy(see Sect. 2.5). Thus, it is used to discuss the nature of phase transition intuitively.

8.1.3 Free Energy in Bragg–Williams Approximation

We introduce another approach to the free energy as a function of order parameter $F(m:T,H)$. Generally, the free energy is given by

$$F = E - TS. \tag{8.22}$$

Here, we give the energy of a state with the magnetization m by

$$E(m) = -\frac{1}{2} NzJm^2, \tag{8.23}$$

and the entropy in the form of the information (Shannon) entropy:

$$S = -k_B \sum_{\sigma_1=\pm1,\sigma_2=\pm1\ldots,\sigma_N=\pm1} P(\sigma_1, \sigma_2 \ldots, \sigma_N) \ln P(\sigma_1, \sigma_2 \ldots, \sigma_N). \tag{8.24}$$

Now, we approximate the probability of spin states $P(\sigma_1, \sigma_2 \ldots, \sigma_N)$ by a product of independent one-spin probabilities $\{p(\sigma_i)\}$:

$$P(\sigma_1, \sigma_2 \cdots, \sigma_N) = \prod_{i=1}^{N} p(\sigma_i). \tag{8.25}$$

Then, entropy is given by

$$\begin{aligned} S &= -k_{\rm B} \sum_{\sigma_1=\pm1,\sigma_2=\pm1\ldots,\sigma_N=\pm1} P(\sigma_1, \sigma_2 \ldots, \sigma_N) \ln P(\sigma_1, \sigma_2 \ldots, \sigma_N) \\ &= -k_{\rm B} N \left(p(1) \ln p(1) + p(-1) \ln p(-1) \right). \end{aligned} \tag{8.26}$$

Because one-spin probabilities $\{p(\sigma_i)\}$ satisfy the relations:

$$p(1) + p(-1) = 1, \quad p(1) - p(-1) = m, \tag{8.27}$$

they are expressed by m as

$$p(1) = \frac{1+m}{2}, \quad p(-1) = \frac{1-m}{2}. \tag{8.28}$$

Substituting these to the entropy, the free energy is given by

$$F_{\rm BW}(m) = -\frac{JzN}{2} m^2 + k_{\rm B} T N \left(\frac{1+m}{2} \ln \frac{1+m}{2} + \frac{1-m}{2} \ln \frac{1-m}{2} \right), \tag{8.29}$$

which is called the 'Bragg–Williams free energy'. The extrema of the free energy are given by

$$\frac{\partial F_{\rm BW}(m)}{\partial m} = 0 \rightarrow Jzm + k_{\rm B} T \frac{1}{2} \ln \left(\frac{1+m}{1-m} \right) = 0. \tag{8.30}$$

This is rewritten as

$$Jzm = k_{\rm B} T \frac{1}{2} \ln \left(\frac{1+m}{1-m} \right) \rightarrow \exp(\beta Jzm) = \frac{1+m}{1-m} \rightarrow m = \tanh(\beta Jzm), \tag{8.31}$$

which agrees with (8.4). Therefore, $F_{\rm BW}(m)$ is also regarded as a free energy of a function of magnetization. This idea is used in various fields. For example, the so-called Slichter–Drickamer model [179] is based on the same idea of the present approximation.

As long as we concern the extrema of the free energy, which give the equilibrium thermodynamic quantities, $F_{\rm BW}(m)$ and $F(m:T,H)$ given in (8.20) give the same results. However, forms of the functions are different below $T_{\rm C}$ as depicted in Fig. 8.2.

There, for $k_B T = 3.7J < k_B T_C$, the dashed line shows $F_{BW}(m)$ and the bold line gives that given in (8.20). The reason for this difference is that $m \neq \tanh(\beta J z m)$ except at the solutions of the self-consistent equation. Indeed, $F_{BW}(m)$ and (8.20) give the same curve at $k_B T = 6J$ and $k_B T = T_C = 4J$, because $m = \tanh(\beta J z m) = 0$.

Thus, when we use a free energy function of magnetization as the potential energy of the order parameter, we should keep this fact in mind. We cannot say which is better. The choice depends on the situations which we concern.

One may consider that the above-mentioned free energies are not for a concrete microscopic model. In the next subsection, we study the free energy functions for a system with long-range interaction which is a model given by an explicit Hamiltonian. We find that it also gives the same free energies as those of mean-field results in the limit $N \to \infty$.

8.1.4 Free Energy of the Long-Range Interaction Model (Husimi–Temperley Model)

The model in which all the spins interact with each other is a model of long-range interaction model, and is called the 'Husimi–Temperley model':

$$\mathcal{H}_{HT} = -\frac{Jz}{2N}\sum_{i=1}^{N}\sum_{j=1}^{N}\sigma_i\sigma_j - H\sum_{i=1}^{N}\sigma_i. \tag{8.32}$$

This model is obtained by substituting $m = \sum_i^N \sigma_i / N$ in the mean-field Hamiltonian:

$$\mathcal{H}_{HT} = -\frac{Jz}{2N}M^2 - HM, \quad M = \sum_{i=1}^{N}\sigma_i. \tag{8.33}$$

8.1.4.1 Free Energy Obtained by an Estimation of the Density State with Stirling's Relation

Let the number of up-spins and down-spins be N_+ and N_-, respectively, which are related to the magnetization as

$$N = N_+ + N_-, \quad M = N_+ - N_- \to N_+ = \frac{N+M}{2}, \quad N_- = \frac{N-M}{2}. \tag{8.34}$$

The entropy is given by

$$S = k_B \ln W, \quad W(N_+) = {}_N C_{N_+} \quad \text{or} \quad W(m) = {}_N C_{(N+M)/2}, \tag{8.35}$$

and the free energy per spin $f(m)$ is given by

$$f(m) = -\frac{Jz}{2}m^2 - Hm - k_B T \ln W(m), \tag{8.36}$$

where

$${}_N C_{(N+M)/2} = \frac{N!}{\left(\frac{N+M}{2}!\right)\left(\frac{N-M}{2}!\right)}. \tag{8.37}$$

Up to here, no approximation was used.

Now, we estimate the entropy by using Stirling's relation for large N:

$$\ln N! \simeq N \ln N - N. \tag{8.38}$$

With this approximation, free energy has the following form:

$$f(m) = -\frac{Jz}{2}m^2 - Hm - \frac{1}{\beta}\left(\frac{1+m}{2}\ln\frac{1+m}{2} + \frac{1-m}{2}\ln\frac{1-m}{2}\right), \tag{8.39}$$

which coincides with the free energy in the Bragg–Williams approximation (8.29).

8.1.4.2 Free Energy Obtained by an Estimation of the Partition Function with the Auxiliary Field Method

Using the relation

$$\frac{1}{\sqrt{2\pi}}\int_{-\infty}^{\infty} e^{-\frac{x^2}{2}+ax}dx = e^{\frac{1}{2}a^2}, \tag{8.40}$$

the partition function is expressed as

$$Z = \mathrm{Tr} e^{\frac{\beta Jz}{2N}M^2+\beta HM} = \mathrm{Tr}\frac{1}{\sqrt{2\pi}}\int_{-\infty}^{\infty} e^{-\frac{x^2}{2}+(x\sqrt{\frac{\beta Jz}{N}}+\beta H)\sum_{i=1}^{N}\sigma_i}dx$$

$$= \frac{1}{\sqrt{2\pi}}\int_{-\infty}^{\infty} e^{-\frac{x^2}{2}+N\ln\left(2\cosh(x\sqrt{\frac{\beta Jz}{N}}+\beta H)\right)}dx. \tag{8.41}$$

This x is called the 'auxiliary field'. Using the variable $x = \sqrt{N}y$, it is

$$Z = \mathrm{Tr}\frac{1}{\sqrt{2N\pi}}\int_{-\infty}^{\infty} e^{-N(\frac{y^2}{2}+\ln[2\cosh(\sqrt{\beta Jz}y+\beta H)])}dy. \tag{8.42}$$

Analyzing the integral by the saddle point method, we have

$$-\beta f(y) = -\frac{1}{2}y^2 + \ln 2\cosh(\sqrt{\beta J z}y + \beta H). \tag{8.43}$$

Changing the variables as

$$y = \sqrt{\beta J z m}, \tag{8.44}$$

we have

$$-\beta f(m) = -\frac{1}{2}\beta J z m^2 + \ln[2\cosh(\beta J z m + \beta H)], \tag{8.45}$$

which agrees with (8.20).

We have estimated the free energy of the long-range interaction model by two methods, which give (8.29) and (8.20). They are different, although both give the same free energies at the extrema. This is due to the fact that the saddle point approximation is good for the minimum point in the limit $N \to \infty$ but does not guarantee values of $f(m)$ apart from the minimum point.

8.1.5 Free Energy as a Variational Function

We may obtain the free energy $f(m)$ as a variational function for the true free energy as explained in Feynman's textbook [180]. Let us express the Hamiltonian in the following form:

$$\mathcal{H} = \mathcal{H}_{\text{MF}}(m) + (\mathcal{H} - \mathcal{H}_{\text{MF}}(m)), \tag{8.46}$$

where $\mathcal{H}_{\text{MF}}(m)$ is given by (8.11). Then, the partition function is given as

$$Z = \text{Tr}e^{-\beta(\mathcal{H}_{\text{MF}}(m)+(\mathcal{H}-\mathcal{H}_{\text{MF}}(m))} = \frac{\text{Tr}e^{-\beta(\mathcal{H}_{\text{MF}}(m)+(\mathcal{H}-\mathcal{H}_{\text{MF}}(m))}}{Z_{\text{MF}}} \times Z_{\text{MF}}. \tag{8.47}$$

Then, the free energy is given in the form

$$F = F_{\text{MF}}(m) - k_{\text{B}}T \ln\langle e^{-\beta(\mathcal{H}-\mathcal{H}_{\text{MF}}(m))}\rangle_{\text{MF}}, \tag{8.48}$$

where $F_{\text{MF}}(m)$ is given by (8.12). Here, we denote the thermal average in the system $\mathcal{H}_{\text{MF}}(m)$ by

$$\langle\cdots\rangle_{\text{MF}} \equiv \frac{\text{Tr}\cdots e^{-\beta(\mathcal{H}_{\text{MF}}(m))}}{Z_{\text{MF}}}. \tag{8.49}$$

Up to here, there is no approximation.

Now, we use the so-called Bogoliubov inequality [180, 181], and have the relation

$$F \le F_{\text{MF}}(m) + \langle\mathcal{H} - \mathcal{H}_{\text{MF}}(m)\rangle_{\text{MF}} \equiv \mathcal{F}_{\text{MF}}(m). \tag{8.50}$$

We regard this relation as a variational estimation of free energy. If we compare this relation with (8.12), we find that $\langle \mathcal{H}_{\rm MF}(m) - \mathcal{H} \rangle_{\rm MF}$ is added to compensate the double count of bonds in $\mathcal{H}_{\rm MF}$.

The second term can be calculated easily because of the independence of each spin, and we have

$$\mathcal{F}_{\rm MF}(m) = -k_{\rm B}TN \ln(2\cosh(z\beta Jm + \beta H))$$

$$- \frac{zJN}{2} \tanh^2(z\beta Jm + \beta H) + zJNm \tanh(z\beta Jm + \beta H). \tag{8.51}$$

The extrema of $\mathcal{F}_{\rm MF}$ are given by

$$\frac{\partial \mathcal{F}_{\rm MF}(m)}{\partial m} = 0, \tag{8.52}$$

which gives

$$-zJNt + \frac{zJNz\beta J}{2} 2t(1-t^2) + zJNt + zJNm\beta Jz(1-t^2)$$

$$-zJmN \tanh(z\beta Jm + \beta H) + \frac{zJN}{2} 2\tanh(z\beta Jm + \beta H)(1 - \tanh^2(z\beta Jm + \beta H))$$

$$+\frac{zJN}{2} m(1 - \tanh^2(z\beta Jm + \beta H)) + \frac{zJN}{2} \tanh(z\beta Jm + \beta H)$$

$$= z^2 J^2 \beta N(1-t^2)(m-t) = 0, \tag{8.53}$$

where we abbreviated $t \equiv \tanh(z\beta Jm + \beta H)$. Because $(1-t^2) > 0$ at finite temperatures, $m = t$ gives the same self-consistent equation (8.4). But, (8.51) is different from either of (8.39) or (8.12) except at points of the solution of the self-consistent equations.

8.2 Equation of Stochastic Processes

Here, we give a brief explanation of the methods of the stochastic process. In thermodynamics, the second law says that the heat flows from the high-temperature side to the low-temperature side. However, there is no explicit equation of motion. Thus, we need some ideas to treat the effect of the heat on the dynamics. As the naivest approach, we may adopt the equation of motion of mechanics including all the degrees of freedom. This is called 'molecular dynamics'. There, no concept of heat is used, but the effect of heat is automatically taken into account. But in many cases of realistic problems, it is not practical because of the huge degrees of freedom. Then, we need to introduce some stochastic approaches, in which the total system is separated into a system which we are interested in and others which are treated as an environment (or thermal bath).

8.2.1 Master Equation and Fokker–Planck Equation

One of the stochastic approaches is a method to use 'transition probability'. The transition probability $w_{i\to j}$ is the probability that the state i changes to j per unit time. The time evolution of the probability of the state i, $P(i)$, in a period Δt, is given by

$$P(i, t+\Delta t) = P(i,t) - \sum_{i\neq j} P(i,t) w_{i\to j}\Delta t + \sum_{i\neq j} P(j,t) w_{j\to i}\Delta t, \tag{8.54}$$

where the second term of the right-hand side is the flow-out probability from the state i to the state j, and the third one is the flow-in probability from j to i. We assume that the transition probability does not depend on the history of the process, i.e., how the present probability has evolved so far. The process satisfying this assumption is called the Markov process. This equation is called the 'master equation'. Here, we introduce a vector expression of $P(i,t)$ as

$$\boldsymbol{P}(t) = \begin{pmatrix} P(1,t) \\ P(2,t) \\ \vdots \\ P(\mathcal{M},t) \end{pmatrix}, \tag{8.55}$$

where $\mathcal{M}$ is the number of states in the system. For the Ising model with N spins,

$$\mathcal{M} = 2^N. \tag{8.56}$$

Using the vector expression (8.55), (8.54) is expressed as

$$\boldsymbol{P}(t+\Delta t) = \mathcal{L}\boldsymbol{P}(t). \tag{8.57}$$

For the master equation (8.54), the time-evolution operator $\mathcal{L}$ is a matrix of $\mathcal{M} \times \mathcal{M}(= 2^N \times 2^N)$ in contrast to the transfer matrix for the equilibrium state which is $2^L \times 2^L$. This $\mathcal{M} \times \mathcal{M}$ matrix $\mathcal{L}$ is called the time-evolution operator, whose matrix element is given by

$$\mathcal{L}_{ij} = \omega_{j\to i}\Delta t \geq 0, \quad i \neq j \tag{8.58}$$

and the diagonal element $\mathcal{L}_{ii}$ is the probability to stay in the same state and it is given by

$$\mathcal{L}_{ii} = 1 - \sum_{j \neq i} \omega_{i \to j} \Delta t, \tag{8.59}$$

which must be nonnegative. We must choose Δt to satisfy this condition. The transition probability $\omega_{i \to j}$ is given for the unit time, and the transition probability in the time interval Δt is $\omega_{i \to j} \Delta t$.

With this choice of the transition probabilities, the following relation trivially holds,

$$\sum_i \mathcal{L}_{ij} = 1. \tag{8.60}$$

This relation guarantees the conservation of the probability

$$\sum_i P(i, 0) = 1 \to \sum_i P(i, t) = 1. \tag{8.61}$$

A matrix which satisfies this condition is called a 'probability matrix'.

Repeating this operation

$$\boldsymbol{P}(t + 2\Delta t) = \mathcal{L}^2 \boldsymbol{P}(t), \tag{8.62}$$

the following properties hold:

$$\mathcal{L}(2\Delta t) = \mathcal{L}(\Delta t)^2, \quad (\mathcal{L}(2\Delta t))_{ij} = \sum_k (\mathcal{L}(\Delta t))_{ik} (\mathcal{L}(\Delta t))_{kj}, \tag{8.63}$$

and for general $t = n\Delta t$ and $s = m\Delta t$, the following property is satisfied:

$$(\mathcal{L}((n+m)\Delta t))_{ij} = \sum_k (\mathcal{L}(n\Delta t))_{ik} (\mathcal{L}(m\Delta t))_{kj} \to (\mathcal{L}(t+s))_{ij} = \sum_k (\mathcal{L}(t))_{ik} (\mathcal{L}(s))_{kj}. \tag{8.64}$$

This relation is called the 'Chapman–Kolmogorov equation', and sometimes it is regarded as a definition of the Markov process.

8.2.2 Master Equation in Differential Form

In the usual modeling of a processes, properties in the limit $\Delta t \to 0$ is to be obtained. Thus, we need to choose a small Δt below which the obtained result does not depend on Δt. To study the case of $\Delta t \to 0$, the differential form of the master equation is convenient.

The time-evolution equation (8.54) is also expressed in a differential form as

$$\frac{\partial P(i,t)}{\partial t} = \lim_{\Delta t \to 0} \frac{P(i,t+\Delta t) - P(i,t)}{\Delta t} = \sum_{i \neq j} P(i,t) w_{i \to j} + \sum_{i \neq j} P(j,t) w_{j \to i}, \tag{8.65}$$

which is expressed as

$$\frac{d}{dt} \boldsymbol{P}(t) = L \boldsymbol{P}(t), \quad L = \frac{\mathcal{L} - 1}{\Delta t}. \tag{8.66}$$

The eigenvectors of L are the same as those of $\mathcal{L}$ (8.57), and the eigenvalues of L

$$L|\phi_m\rangle = \lambda_m |\phi_m\rangle \tag{8.67}$$

are given by

$$\lambda_k = \frac{\hat{\lambda}_k - 1}{\Delta t}. \tag{8.68}$$

It should be noted that for the stationary state $\boldsymbol{P}_{\rm st} = |\phi_1\rangle$

$$\lambda_1 = 0. \tag{8.69}$$

The time-evolution function $\mathcal{L}$ is given by

$$\mathcal{L}(\Delta t) = e^{-L\Delta t}. \tag{8.70}$$

8.2.3 Symmetrization of the Time-Evolution Operator

The time-evolution operator L (or $\mathcal{L}$) is not a symmetric matrix, i.e., $L_{ij} \neq L_{ji}$. However, we may have a symmetrized time-evolution operator. Noting the detailed balance:

$$w_{i \to j} P_{\rm eq}(i) = w_{j \to i} P_{\rm eq}(j) \to L_{ji} P_{\rm eq}(i) = L_{ij} P_{\rm eq}(j), \tag{8.71}$$

let us consider a matrix

$$L_{\rm sym} \equiv (\hat{P}_{\rm eq})^{-1/2} L (\hat{P}_{\rm eq})^{1/2}, \tag{8.72}$$

where

$$\hat{P}_{\rm eq} = \begin{pmatrix} e^{-\beta E_1} & 0 & 0 & \cdots, 0 & \\ 0 & e^{-\beta E_2} & 0 & \cdots, 0 & \\ & & \ddots & & \\ 0 & \cdots & 0 & 0 & e^{-\beta E_{2^N}} \end{pmatrix}. \tag{8.73}$$

The ij component is

$$\left((\hat{P}_{\rm eq})^{-1/2}L(\hat{P}_{\rm eq})^{1/2}\right)_{ij} = (\hat{P}_{\rm eq})_{ii}^{-1/2}L_{ij}(\hat{P}_{\rm eq})_{jj}^{1/2} = P_{\rm eq}(i)^{-1/2}L_{ji}P_{\rm eq}(i)P_{\rm eq}(j)^{-1}P_{\rm eq}(j)^{1/2}$$

$$= (\hat{P}_{\rm eq})_{jj}^{-1/2}L_{ji}(\hat{P}_{\rm eq})_{ii}^{1/2} = \left((\hat{P}_{\rm eq})^{-1/2}L(\hat{P}_{\rm eq})^{1/2}\right)_{ji}, \tag{8.74}$$

and we find that the matrix $L_{\rm sym}$ is a symmetric matrix. The matrix $(\hat{P}_{\rm eq})^{-1/2}L(\hat{P}_{\rm eq})^{1/2}$ has the same eigenvalues as those of L, and this form is convenient for numerical calculation.

8.2.4 Master Equation for Continuous Variable

In cases of systems with continuous states, the states are denoted by a continuous variable x, and the transition probability is given by $W(x; r) \equiv w_{x\to x+r}$. Replacing the sum by an integration, the master equation in this case is given by

$$\frac{\partial P(i,t)}{\partial t} = \int W(x-r;r)P(x-r,r)dr - P(x,t)\int W(x;r)dr. \tag{8.75}$$

Expanding the equation in powers of r, and introducing moments of $W(x;r)$:

$$a_j(x) = \int_{-\infty}^{\infty} r^j W(x;r)dr, \tag{8.76}$$

the master equation is given as

$$\frac{\partial P(i,t)}{\partial t} = -\frac{\partial}{\partial x}(a_1(x)P(x,t)) + \frac{1}{2}\frac{\partial^2}{\partial x^2}(a_2(x)P(x,t)) + \cdots, \tag{8.77}$$

which is called the 'Kramers–Moyal equation'. In the case where the higher order terms than the second one can be ignored, the equation is called the Fokker–Planck equation.

8.2.5 Brownian Motion

Another method uses a random variable representing an external noise. The most idealized noise is the white-Gaussian noise which is used in the analysis of the Brownian motion. The motion of a particle in a random force $\xi(t)$ is given by

$$m\frac{d^2x}{dt^2} = \xi(t). \tag{8.78}$$

Hereafter, we set the mass m to be unity ($m = 1$) for simplicity. The change of the velocity $v = dx/dt$ is given by

$$\frac{dv}{dt} = \xi(t). \quad v(t) = \int_0^t \xi(s)ds + v(0). \tag{8.79}$$

If we assume that the noise fluctuates very fast and its average is zero, i.e.,

$$\langle \xi(t) \rangle = 0, \quad \langle \xi(t)\xi(t') \rangle = 2D\delta(t - t'), \tag{8.80}$$

where $\langle \cdots \rangle$ denotes the average over distribution of the random noise, and D represents the strength of the noise. Moreover, we assume that higher cumulants of $\xi(t)$ are zero. When $\xi(t)$ satisfies these conditions, the noise is called 'white-Gaussian noise', and the quantity

$$W(t) \equiv \int_0^t \xi(s)ds \tag{8.81}$$

is called the 'Wiener process', which has the following properties:

$$\langle W(t) \rangle = 0, \quad \langle W(t)^2 \rangle = 2Dt, \qquad \langle W(t)W(t') \rangle = 2D\min(t, t'). \tag{8.82}$$

The probability of $W(t)$ is given by

$$P(W(t)) = \frac{1}{\sqrt{4\pi Dt}} e^{-W(t)^2/4Dt}. \tag{8.83}$$

8.2.5.1 Langevin Equation

The equation of motion including a random variable is called the 'Langevin equation'. For example, a motion in friction proportional to its velocity is given by

$$m\frac{dv}{dt} = -kv + \xi(t), \tag{8.84}$$

where k is a coefficient of the friction. This process is called the ' Ornstein–Uhlenbeck process', which gives a model for the Brownian motion. This is formally a linear equation of v, and the solution is given by

$$v(t) = v(0)e^{-\frac{k}{m}t} + \frac{1}{m}\int_0^t e^{-\frac{k}{m}(t-t')}\xi(t')dt'. \tag{8.85}$$

Using the properties (8.80), the average is

$$\langle v(t)\rangle = v(0)e^{-\frac{k}{m}t}, \tag{8.86}$$

and the variance is

$$\sigma(t)^2 \equiv \langle (v(t) - \langle v(t)\rangle)^2\rangle = \left\langle \left(\frac{1}{m}\int_0^t e^{-\frac{k}{m}(t-t')}\xi(t')dt'\right)\left(\frac{1}{m}\int_0^t e^{-\frac{k}{m}(t-s')}\xi(s')ds'\right)\right\rangle$$

$$= 2D\frac{1}{m^2}e^{-2\frac{k}{m}t}\int_0^t e^{-2\frac{k}{m}t'}dt' = \frac{D}{km}\left(1 - e^{-2\frac{k}{m}t}\right). \tag{8.87}$$

Because $\xi(t)$ is the Gaussian distribution, distributions of any sum of $\{\xi(t_j)\}$, $j = 1, 2, \ldots$ are Gaussian. Thus, the probability of $v(t)$ for the initial value $v(0)$ is given by a Gaussian form, and it is determined by $\langle v(t)\rangle$ and $\sigma(t)^2$ as

$$P(v(t), v(0)) = \sqrt{\frac{1}{2\pi\sigma(t)^2}}\exp\left(-\frac{1}{2}\frac{(v(t) - \langle v(t)\rangle)^2}{\sigma(t)^2}\right)$$

$$= \sqrt{\frac{km}{2\pi D\left(1 - e^{-2\frac{k}{m}t}\right)}}\exp\left(-\frac{km}{2}\frac{(v(t) - \langle v(t)\rangle)^2}{D\left(1 - e^{-2\frac{k}{m}t}\right)}\right). \tag{8.88}$$

For large t, $e^{-2(k/m)t} \simeq 0$ and $\langle v(t)\rangle \simeq 0$. Thus,

$$\lim_{t\to\infty} P(v(t), v(0)) = \sqrt{\frac{km}{2\pi D}}\exp\left(-\frac{km}{2}\frac{v(t)^2}{D}\right), \tag{8.89}$$

which is independent of the initial value $v(0)$. If we identify this distribution as the thermal equilibrium distribution

$$P(v) = \sqrt{\frac{m}{2\pi k_B T}}\exp\left(-\frac{m}{2}\frac{v^2}{k_B T}\right), \tag{8.90}$$

we have the relation

$$\frac{km}{2}\frac{1}{D} = \frac{m}{2}\frac{1}{k_B T} \to D = kk_B T. \tag{8.91}$$

Here, D is the strength of the fluctuation force given by (8.80). This is one of the Einstein relations. If we study the variance of the position $x(t)$, and perform similar calculations (not shown but straightforward), for a large t, we have

$$\langle (x(t) - \langle x(t) \rangle)^2 \rangle \simeq 2\frac{D}{k^2}t. \tag{8.92}$$

If we define the diffusion constant of the position as

$$\langle (x(t) - \langle x(t) \rangle)^2 = 2D_{\mathrm{diff}}t, \tag{8.93}$$

we have the relation

$$D_{\mathrm{diff}} = \frac{D}{k^2} \to D_{\mathrm{diff}} = \frac{k_{\mathrm{B}}T}{k}, \tag{8.94}$$

which is also one of the Einstein relations.

For small $t = \Delta t$,

$$\left(1 - e^{-2\frac{k}{m}\Delta t}\right) \simeq 2\frac{k}{m}\Delta t \tag{8.95}$$

and thus

$$P(v(\Delta t), v(0)) \simeq \sqrt{\frac{m^2}{4\pi D\Delta t}} \exp\left(-\frac{m^2}{4}\frac{(v(t) - \langle v(\Delta t)\rangle)^2}{D\Delta t}\right), \tag{8.96}$$

which is called the Gaussian propagator for the diffusion process for the period Δt.

The general form of the Langevin equation for a quantity x is given by

$$\frac{dx}{dt} = -f(x) + g(x)\xi(t). \tag{8.97}$$

Because the amplitude of $W(\Delta t)$ is not proportional to Δt but is proportional to $\sqrt{t}$, the interpretation of the differential equation (8.97) has ambiguity. That is,

$$x(\Delta t) = x(0) - \int_0^{\Delta t} (f(x(t')) + g(x(t'))\xi(t'))dt' \tag{8.98}$$

is not uniquely defined. Namely, we may use either the definition

$$x(\Delta t) = x(0) - f(x(t')\Delta t + g(x(0))W(\Delta t) \tag{8.99}$$

which is called the Ito formula, or

$$x(\Delta t) = x(0) - f(x(t')\Delta t + g\left(\frac{x(0) + x(\Delta t)}{2}\right)W(\Delta t) \tag{8.100}$$

which is called the Stratonovich formula. Both formulae give different results [182]. Neither of them is wrong, but they are just different definitions. It is known that the Stratonovich formula gives the same result as the solution when we assume $\xi(t)$ as a normal continuous function of t and then take the limit the autocorrelation time $\to 0$. In this way, the Stratonovich formula is easily understood, but the Ito formula is easier in numerical calculation. In the following, we give the Langevin equation in the Stratonovich interpretation. It is known that the equation (8.97) given by the Stratonovich interpretation is transformed into the equation given by the Stratonovich interpretation by the following replacement:

$$f(x(t))_{\text{Stratonovich}} \to f(x(t))_{\text{Ito}} = f(x(t))_{\text{Stratonovich}} - Dg(x)\frac{dg(x)}{dx}, \tag{8.101}$$

which called the 'Ito–Stratonovich transformation'.

8.2.5.2 Stochastic Liouville Equation

The equation of continuity of the distribution of x is given by

$$\frac{\partial f(x,t)}{\partial t} + \frac{\partial}{\partial x}(vf(x,t)) = 0, \quad v = \frac{dx}{dt}. \tag{8.102}$$

Substituting this relation to (8.97), we have

$$\frac{\partial f(x,t)}{\partial t} + \frac{\partial}{\partial x}((h(x) + g(x)\xi(t))f(x,t)) = 0, \quad v = \frac{dx}{dt}, \tag{8.103}$$

which is called the 'Stochastic Liouville equation'. The probability function $P(x, t)$ is given by the average of $f(x)$ over the random noise$\xi(t)$:

$$P(x,t) = \langle f(x,t)\rangle. \tag{8.104}$$

Let us define operations:

$$L_0 = -\frac{\partial}{\partial x}h(x), \quad L_1 = -\frac{\partial}{\partial x}(g(x)\xi(t)). \tag{8.105}$$

Then, (8.97) is given as

$$\frac{\partial}{\partial t}f = (L_0 + L_1)f. \tag{8.106}$$

Here, we use a kind of interaction representation for L_0, and define $\mathcal{L}_0(t)$ as the time-evolution operator for L_0:

$$\frac{\partial}{\partial t}\rho_0 = L_0\rho_0, \quad \rho_0(t) = \mathcal{L}_0(t)\rho_0(0). \tag{8.107}$$

$\mathcal{L}_0(t)$ satisfies the relations:

$$\frac{\partial}{\partial t}\mathcal{L}_0 = L_0\mathcal{L}_0, \quad \mathcal{L}_0(t) = Te^{L_0 t}. \tag{8.108}$$

We express *f* by

$$f = \mathcal{L}_0(t)\rho_1, \tag{8.109}$$

and substitute it in (8.106). Then, we have

$$\frac{\partial}{\partial t}\mathcal{L}_0(t)\rho_1 = \left(\frac{\partial}{\partial t}\mathcal{L}_0(t)\right)\rho_1 + \mathcal{L}_0(t)\frac{\partial}{\partial t}\rho_1 = (L_0 + L_1)\mathcal{L}_0(t)\rho_1, \tag{8.110}$$

and thus

$$\mathcal{L}_0\frac{\partial}{\partial t}\rho_1 = L_1\mathcal{L}_0\rho_1. \tag{8.111}$$

From this,

$$\frac{\partial}{\partial t}\rho_1 = \mathcal{L}_0^{-1}L_1\mathcal{L}_0\rho_1 = \xi(t)\mathcal{L}_0^{-1}g(x)\mathcal{L}_0\rho_1 \equiv \hat{L}_1\rho_1. \tag{8.112}$$

Thus,

$$\rho_1(t) = T\exp\left[\int_0^t \hat{L}(s)ds\right]\rho_1(0). \tag{8.113}$$

Averaging *f* over $\xi(t)$, $P(x, t)$ is given by

$$P(x,t) = \mathcal{L}_0(t)\langle T\exp\left[\int_0^t \hat{L}(s)ds\right]\rho_1(0)\rangle. \tag{8.114}$$

Because $\langle\hat{L}_1(t)\rangle = 0$, using a general property of the Gauss distribution, we have

$$\rho_1(t) = T\exp\left[\frac{1}{2}\left\langle\int_0^t \hat{L}(s)ds\int_0^t \hat{L}(s')ds'\right\rangle\right]\rho_1(0). \tag{8.115}$$

Using the relation $\langle\xi(s)\xi(s')\rangle = 2D\delta(s - s')$, we have

$$\left\langle\frac{1}{2}\int_0^t \hat{L}(s)ds\int_0^t \hat{L}(s')ds'\right\rangle = D\int_0^t \mathcal{L}_0^{-1}\frac{\partial}{\partial x}g(x(s))\frac{\partial}{\partial x}g(x(s))\mathcal{L}_0 ds. \tag{8.116}$$

Thus, $P(x, t)$ is given by

$$P(x,t) = \mathcal{L}_0(t)\left\langle T\exp\left[D\int_0^t \mathcal{L}_0^{-1}\frac{\partial}{\partial x}g(x(s))\frac{\partial}{\partial x}g(x(s))\mathcal{L}_0 ds\right]\rho_1(0)\right\rangle. \tag{8.117}$$

Differentiating this by t noting that $\frac{\partial}{\partial t}\mathcal{L}_0 = L_0\mathcal{L}_0 = -\frac{\partial}{\partial x}h(x)\mathcal{L}_0$, the time-evolution equation of $P(x,t)$ is obtained as

$$\frac{\partial}{\partial t}P(x,t) = \left(-\frac{\partial}{\partial x}h(x) + D\frac{\partial}{\partial x}g(x)\frac{\partial}{\partial x}g(x)\right)P(x,t) \tag{8.118}$$

in a form of the Fokker–Planck equation. It should be noted that the second term has a different form from the standard form of the Fokker–Planck equation when $g(x)$ is not a constant.

8.3 Landau–Zener Scattering

As a typical model of population dynamics at an avoided level crossing point, the following two-level system is often used:

$$\mathcal{H} = -H(t)\sigma^z - \Gamma\sigma^x, \quad \sigma^z = \begin{pmatrix} 1 & 0 \\ 0 & -1 \end{pmatrix}, \quad \sigma^x = \begin{pmatrix} 0 & 1 \\ 1 & 0 \end{pmatrix}, \tag{8.119}$$

in which the field $H(t)$ is swept with a velocity c:

$$H(t) = ct + H_0. \tag{8.120}$$

The population dynamics of the model is called the ‘Landau–Zener scattering’ [84–86]. Applications to magnetics systems also have been studied [87–90].

The Schrödinger equation is given by

$$i\hbar\frac{\partial}{\partial t}\begin{pmatrix} a \\ b \end{pmatrix} = \begin{pmatrix} -H(t) & -\Gamma \\ -\Gamma & H(t) \end{pmatrix}\begin{pmatrix} a \\ b \end{pmatrix}. \tag{8.121}$$

The eigenvalues of the Hamiltonian at a given time t are

$$E_\pm = \pm\sqrt{H(t)^2 + \Gamma^2}, \tag{8.122}$$

which shows an avoided level structure with an energy gap $\Delta E = 2\Gamma$ for $H_0 = 0$ is depicted in Fig. 8.3.

The constant field H_0 tilts the avoided level structure, but the effect of H_0 can be taken into account by using the transformation:

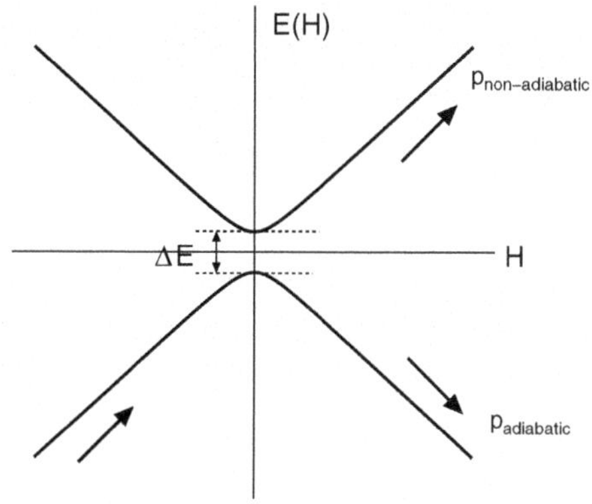

Fig. 8.3 Landau–Zener scattering at an avoided level crossing point

$$\begin{pmatrix} a \\ b \end{pmatrix} \to \begin{pmatrix} a \\ b \end{pmatrix} e^{-iH_0 t/\hbar}. \quad (8.123)$$

Thus, we hereafter set $H_0 = 0$. In the following, we follow Zener's paper [85], where the field is swept from a negative side to the positive side, i.e., a negative velocity. In the following calculation, we put $\hbar = 1$.

At $t = -\infty$, $H = ct = -\infty$ and the state is

$$|t = -\infty\rangle = \begin{pmatrix} 0 \\ 1 \end{pmatrix}. \quad (8.124)$$

After the crossing point, it becomes

$$|t = \infty\rangle = \begin{pmatrix} a(\infty) \\ b(\infty) \end{pmatrix}. \quad (8.125)$$

Eliminating b in (8.121), we have

$$\frac{d^2a}{dt^2} + 2ci\frac{da}{dt} + \Gamma^2 a = 0. \quad (8.126)$$

Changing the parameter

$$a = e^{-ict} U_1, \quad (8.127)$$

(8.126) is written as

$$\frac{d^2 U_1}{dt^2} + \left(\Gamma^2 - ic + c^2 t^2\right) U_1 = 0. \quad (8.128)$$

By using new variables [85]

$$z = \sqrt{2c}e^{-i\pi/4}t, \quad n = i\frac{\Gamma^2}{2c}, \tag{8.129}$$

Equation (8.128) is rewritten in the form of the Weber equation:

$$\frac{d^2U_1}{dz^2} + \left(n + \frac{1}{2} - \frac{z^2}{4}\right)U_1 = 0, \tag{8.130}$$

and the solution is given by the Weber function [183]. By making use of the asymptotic form of the Weber function, the solution satisfying the initial condition (8.124) is found, and the scattering probabilities are obtained, including $\hbar$, as

$$|a(\infty)|^2 = 1 - \exp\left(\frac{-\pi\Gamma^2}{\hbar c}\right), \quad |b(\infty)|^2 = \exp\left(\frac{-\pi\Gamma^2}{\hbar c}\right), \tag{8.131}$$

where c is the velocity of the Zeeman energy. This formula has been obtained by various papers in the same year 1932 [84–86]. Thus, using the energy gap $\Delta E = 2\Gamma$, the formula is written as

$$|a(\infty)|^2 = 1 - \exp\left(\frac{-\pi(\Delta E)^2}{4\hbar c}\right), \quad |b(\infty)|^2 = \exp\left(\frac{-\pi(\Delta E)^2}{4\hbar c}\right), \tag{8.132}$$

which is called the 'Landau–Zener formula'.

Sweep velocity dependence of magnetization dynamics of the Landau–Zener scattering (8.119) is depicted in Fig. 8.4, where the field is swept from the negative side to the positive side, i.e., $c > 0$. The magnetization is given by the population at the positive branch p_+ and that of negative branch p_-:

$$m = p_+ - p_-. \tag{8.133}$$

In the avoided level crossing region, the state changes largely, and after the scattering the populations of both levels converge to the value given by (8.131). It should be noted that in quantum dynamics, there is no mechanism of transfer between levels except the crossing point. Thus, the population scattered to each level at the crossing points does not relax. In realistic cases, dissipation effects are important as discussed in Sect. 6.8.

It is also noted that the amount of adiabatic transition depends largely on the gap. For example, let us consider a system of a single spin in which the gap is 1GHz, i.e., $\Delta E/\hbar = 10^9$/s. In this system, the Zeeman energy is $E_H = g\mu_B H$ where $\mu_B = 0.927 \times 10^{-23}$ J/T. When the change of magnetic field is given in the unit of tesla/second, c is given by $c = dH(t)/dt = \hbar g\mu_B v$.

As a point where a significant change occurs, we consider the case

$$\frac{\Delta E^2}{\hbar g\mu_B v} = 1, \tag{8.134}$$

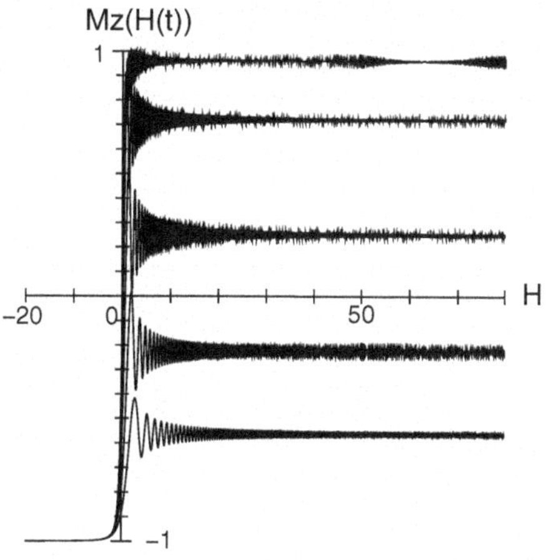

Fig. 8.4 **a** Dynamics of the magnetization for $\Gamma = 0.5$ for field sweeps: $c = 1.6$, 0.8, 0.4, 0.2, and 0.1 from bottom to top

where

$$\exp\left(\frac{-\pi(\Delta E)^2}{4\hbar c}\right) = \exp\left(\frac{-\pi}{4}\right). \tag{8.135}$$

Substituting the values, we have

$$\frac{\Delta E^2}{\hbar v} = \frac{10^{18} \times (1.054 \times 10^{-34})^2}{1.054 \times 10^{-34} \times 0.927 \times 10^{-23} v} = \frac{1.054 \times 10^7}{0.927 v}. \tag{8.136}$$

Thus, for the condition (8.134), we need $v \sim 10^7$ T/s, which is very fast, and the non-adiabatic change is hardly seen in realistic situations. On the other hand, for the gap of 1MHz, i.e., $\Delta E/\hbar = 10^6$, $v \sim 10$ T/s, and also for a much small gap, e.g., $\Delta E/\hbar = 10^3$, $v \sim 10^{-5}$ T/s for which the adiabatic change is hardly seen in realistic situations.

8.4 Quantum Master Equation

The quantum master equation is generally given by the following form:

$$\frac{d\rho}{dt} = -i[\mathcal{H}, \rho] + D[\rho], \tag{8.137}$$

where ρ is the density matrix of the system. The first term of the right-hand side gives the pure quantum dynamics, and $D[\rho]$ gives the dissipation effect. Here, we survey methods to treat the dissipation effects.

8.4.1 Lindblad Type

The Lindblad formula [184, 185] for the master equation is generally given by

$$D[\rho] = -\sum_{A_j} \gamma_j \left(2A_j\rho(t)A_j^\dagger - A_j^\dagger A_j\rho(t) - \rho(t)A_j^\dagger A_j\right), \tag{8.138}$$

where $\gamma_j \geq 0$ and A_j is an operator. This formula is sometimes called the 'Gorini–Kossakowski–Sudarshan–Lindblad equation(GKSL equation)' [186]. This formula guarantees the positivity of probability in time evolutions, and is often used to study formal operation processes in dissipative environments. For the two-level systems, the relaxation to the equilibrium state can be easily realized by choosing parameters as we see below. However, it is generally difficult for this formula to make the stationary state of the time evolution to be the thermal equilibrium state, although some treatment, such as the dressed Lindblad method [131, 185], can recover the property.

For a single spin system, the dissipative term $D[\rho]$ is given by the following form:

$$\begin{aligned} D[\rho] = &-\alpha_+ \left(S^-S^+\rho - 2S^+\rho S^- + \rho S^-S^+\right) \\ &-\alpha_- \left(S^+S^-\rho - 2S^-\rho S^+ + \rho S^+S^-\right) \\ &-\alpha_z \left(S^zS^z\rho - 2S^z\rho S^z + \rho S^zS^z\right). \end{aligned} \tag{8.139}$$

For the case $S = 1/2$, using the matrix representations:

$$S^+ = \begin{pmatrix} 0 & 1 \\ 0 & 0 \end{pmatrix}, \quad S^- = \begin{pmatrix} 0 & 0 \\ 1 & 0 \end{pmatrix}, \quad S^z = \frac{1}{2}\begin{pmatrix} 1 & 0 \\ 0 & -1 \end{pmatrix}, \tag{8.140}$$

the matrices of the terms are given as

$$S^-S^+\rho - 2S^+\rho S^- + \rho S^-S^+ = \begin{pmatrix} -2\rho_{22} & \rho_{12} \\ \rho_{21} & 2\rho_{22} \end{pmatrix} \tag{8.141}$$

$$S^-S^+\rho - 2S^+\rho S^- + \rho S^-S^+ = \begin{pmatrix} 2\rho_{11} & \rho_{12} \\ \rho_{21} & -2\rho_{11} \end{pmatrix} \tag{8.142}$$

and

$$S^zS^z\rho - 2S^z\rho S^z + \rho S^zS^z = \begin{pmatrix} 0 & \rho_{12} \\ \rho_{21} & 0 \end{pmatrix}. \tag{8.143}$$

Thus, (8.139) is written as

$$D[\rho] = -\begin{pmatrix} 2\alpha_-\rho_{11} - 2\alpha_+\rho_{22} & (\alpha_+ + \alpha_- + \alpha_z)\rho_{12} \\ (\alpha_+ + \alpha_- + \alpha_z)\rho_{21} & 2\alpha_+\rho_{22} - 2\alpha_-\rho_{11} \end{pmatrix}. \tag{8.144}$$

At a finite temperature, the state should relax to the canonical distribution, i.e.,

$$\rho_{\text{eq}} = \frac{1}{Z} e^{-\beta \mathcal{H}}, \quad Z = e^{-\beta E_1} + e^{-\beta E_2}, \tag{8.145}$$

whose matrix elements are

$$\rho_{11} = \frac{e^{-\beta E_1}}{Z}, \quad \rho_{22} = \frac{e^{-\beta E_2}}{Z}, \quad \rho_{12} = \rho_{21} = 0, \tag{8.146}$$

where E_1 and E_2 are eigenvalues of the Hamiltonian. To have the above density matrix ρ_{eq} as the stationary state of the time evolution, the parameters need to satisfy $d\rho_{\text{eq}}/dt = 0$. The first term of (8.137) is zero for ρ_{eq}, because $e^{-\beta \mathcal{H}}$ and $\mathcal{H}$ commute. To make $D[\rho] = 0$, the relation:

$$\alpha_- \rho_{11} - \alpha_+ \rho_{22} = \alpha_- e^{-\beta E_1} - \alpha_+ e^{\beta E_2} = 0 \rightarrow \frac{\alpha_-}{\alpha_+} = e^{\beta \Delta E} \tag{8.147}$$

is required, where $\Delta E = E_1 - E_2$. With this relation, the Lindblad equation at a finite temperature is often given in the form:

$$\begin{aligned} D[\rho] = -\kappa & \left(m(\beta) \left(S^- S^+ \rho - 2S^+ \rho S^- + \rho S^- S^+ \right) \right. \\ & \left. -(1 - m(\beta)) \left(S^+ S^- \rho - 2S^- \rho S^+ + \rho S^+ S^- \right) \right) \\ & - \kappa_z \left((S^z)^2 \rho - 2S^z \rho S^z + \rho (S^z)^2 \right), \end{aligned} \tag{8.148}$$

where $m(\beta)$ is the Boltzmann distribution:

$$m(\beta) = \frac{e^{-\beta E_1}}{e^{-\beta E_1} + e^{-\beta E_2}}, \quad 1 - m(\beta) = \frac{e^{-\beta E_2}}{e^{-\beta E_1} + e^{-\beta E_2}}. \tag{8.149}$$

8.4.2 Redfield Type

Quantum dynamics in dissipative environments has been studied by a projection method in which the freedoms of the thermal reservoir are integrated out [187–190]. To formulate such a procedure, we consider the following Hamiltonian of the total system:

$$\mathcal{H}_{\text{tot}} = \mathcal{H}_{\text{S}} + \mathcal{H}_{\text{int}} + \mathcal{H}_{\text{B}}, \tag{8.150}$$

where $\mathcal{H}_{\text{S}}$ denotes a system which we are interested in, $\mathcal{H}_{\text{B}}$ denotes a system of the thermal bath, and $\mathcal{H}_{\text{int}}$ denotes the interaction between the system and the bath. We may take a bath consisting of independent harmonic oscillators [106]:

$$\mathcal{H}_{\text{B}} = \sum_{\alpha} \hbar \omega_{\alpha} \left(b_{\alpha}^{\dagger} b_{\alpha} \right), \tag{8.151}$$

where $b_\alpha^\dagger$ and b_α are the creation and annihilation operators of the αth mode of the harmonic system. We consider the interaction given in the form

$$\mathcal{H}_{\rm int} = \lambda \sum_\alpha \sqrt{\hbar}\gamma_\alpha (b_\alpha^\dagger + b_\alpha) X_\alpha, \tag{8.152}$$

where X_α is an operator in the system and γ_α is a coupling strength to the αth bath operator. The parameter λ is a parameter to control strength of the coupling. For the stability of the interaction part, we need a term of $\lambda'^2 X_\alpha^2 (\lambda' > \lambda)$. But here, we ignore it because λ'^2 may be small or we may regard this term as a part of $\mathcal{H}_{\rm S}$.

The total system is an isolated system, and the dynamics follows quantum mechanics and is given by

$$\frac{\partial \rho_{\rm tot}(t)}{\partial t} = \frac{1}{i\hbar}\left[\mathcal{H}_{\rm tot}, \rho_{\rm tot}(t)\right]. \tag{8.153}$$

Under the condition that the thermal bath is in equilibrium at a temperature T, we adopt the density matrix of the bath as

$$\rho_{\rm B} = \frac{1}{Z_{\rm B}} e^{-\beta\mathcal{H}_{\rm B}}, \quad Z_{\rm B} = {\rm Tr} e^{-\beta\mathcal{H}_{\rm B}}, \tag{8.154}$$

where ${\rm Tr}_{\rm B}$ denotes the sum of all the states of the thermal bath.

Now, we trace out the degrees of freedom of the thermal bath to obtain the density matrix of the system $\rho(t)$

$$\rho(t) = {\rm Tr}_{\rm B}\rho_{\rm tot}(t). \tag{8.155}$$

By expanding (8.153) in a series of λ up to the second order and also adopting the Markovian approximation, we obtain the quantum master equation in the following form [188–190], which is called the Redfield type master equation [191]:

$$\frac{\partial \rho(t)}{\partial t} = \frac{1}{i\hbar}\left[\mathcal{H}_{\rm tot}, \rho(t)\right] - \lambda^2 \Gamma[\rho(t)], \tag{8.156}$$

with the dissipation term

$$\Gamma[\rho(t)] = \frac{1}{\hbar^2}\int_0^\infty dt' \int_{-\infty}^{\infty} e^{i\omega t'}\Phi(\omega)$$

$$\times \left\{ XX(-t')\rho(t) - e^{\beta\hbar\omega} X\rho(t)X(-t') + e^{\beta\hbar\omega}\rho(t)X(-t')X - X(-t')\rho(t)X \right\}, \tag{8.157}$$

where

$$X(-t') = e^{-i\mathcal{H}_{\rm S}t'/\hbar} X e^{i\mathcal{H}_{\rm S}t'/\hbar}, \tag{8.158}$$

and $\Phi(\omega)$ is the Fourier transform of the correlation function $C(t)$ of the bath variables

$$\Phi(\omega) = \frac{1}{2\pi}\int_{-\infty}^{\infty} e^{-i\omega t} C(t)dt. \tag{8.159}$$

The correlation function $C(t)$ is given by

$$\begin{aligned} C(t) &= \hbar\langle \gamma_\alpha(b_\alpha^\dagger + b_\alpha)\gamma_\alpha(b_\alpha(t)^\dagger + b_\alpha(t))\rangle \\ &= \hbar\sum_\alpha \gamma_\alpha^2 \frac{e^{i\omega_\alpha t} + e^{\beta\hbar\omega_\alpha} e^{i\omega_\alpha t}}{e^{\beta\hbar\omega_\alpha} - 1}. \end{aligned} \tag{8.160}$$

We introduce the density of states $D(\omega)$ of the amplitude of $\omega = |\omega_\alpha|$ of the thermal bath, and then $\Phi(\omega)$ is written as

$$\Phi(\omega) = \hbar\gamma(\omega)^2 \frac{D(\omega) - D(-\omega)}{e^{\beta\hbar\omega_\alpha} - 1}, \tag{8.161}$$

where $\gamma(\omega)$ gives amplitude dependence of γ_α. The so-called spectral density is given by

$$I(\omega) = \gamma(\omega)^2 D(\omega). \tag{8.162}$$

For the spectral density, the following form is often used:

$$I(\omega) = I_0\omega^\alpha, \quad \text{for } \omega > 0 \quad \text{and} \quad I(\omega) = 0 \text{ for } \omega \leq 0. \tag{8.163}$$

The cases of $\alpha = 1$, $\alpha < 1$, and $\alpha > 1$ are called 'Ohmic', 'sub-Ohmic', and 'super-Ohmic', respectively. Sometimes, a form with a cutoff frequency ω_c:

$$I(\omega) \sim \omega \times e^{-\omega/\omega_c} \tag{8.164}$$

is used [192].

For the integral from 0 to ∞ in (8.157), we use the relation:

$$\int_0^\infty dt e^{i\omega t} = \pi\delta(\omega) + P\frac{i}{\omega}, \tag{8.165}$$

where P denotes the principal value integral. The imaginary part gives a change of the system Hamiltonian. This contribution to the Hamiltonian is sometimes called the 'Lamb shift'.

Neglecting this imaginary contribution, (8.156) is given by a compact form:

$$\frac{\partial\rho(t)}{\partial t} = \frac{1}{i\hbar}\left[\mathcal{H}_{\text{tot}}, \rho(t)\right] - \frac{\pi\lambda^2}{\hbar}\left([X, R\rho(t)] + [X, R\rho(t)]^\dagger\right), \tag{8.166}$$

where R is a matrix whose matrix element is given as

$$\langle l|R|m\rangle = \frac{\langle l|X|m\rangle}{\hbar} \frac{I\left(\frac{E_l-E_m}{\hbar}\right) - I\left(\frac{E_m-E_l}{\hbar}\right)}{e^{\beta(E_l-E_m)}-1}, \tag{8.167}$$

where $|l\rangle$ and $|m\rangle$ are eigenstates of $\mathcal{H}_S$ for the eigenvalues E_l and E_m, respectively, i.e.,

$$\mathcal{H}_S|l\rangle = E_l|l\rangle. \tag{8.168}$$

8.4.2.1 Stationary State of Redfield Type Master Equation

The canonical distribution $\rho_{\text{eq}} = e^{-\beta\mathcal{H}}/Z$ is the stationary state of the Redfield type master equation. Let us check this property.

Trivially,

$$-i[\mathcal{H}, \rho_{\text{eq}}] = -i\left(\mathcal{H}\rho_{\text{eq}} - \rho_{\text{eq}}\mathcal{H}\right) = 0. \tag{8.169}$$

Now check

$$D[\rho_{\text{eq}}] = -\frac{\pi\lambda^2}{\hbar}\left(XR\rho_{\text{eq}} - R\rho_{\text{eq}}X + \rho_{\text{eq}}R^\dagger X - X\rho_{\text{eq}}R^\dagger\right) = 0. \tag{8.170}$$

Using the eigenstates $\{|l\rangle\}$ of $\mathcal{H}$, the (l, m) component of $D[\rho_{\text{eq}}]$ is

$$D_{lm} = \sum_{ij}\left(X_{li}R_{ij}\rho_{jm} - R_{li}\rho_{ij}X_{jm} - X_{li}\rho_{ij}(R^\dagger)_{jm} + \rho_{li}(R^\dagger)_{ij}X_{jm}\right). \tag{8.171}$$

Because for $\rho_{\text{eq}} = e^{-\beta\mathcal{H}}/Z$,

$$\rho_{jm} = e^{-\beta E_j}\delta_{jm}/Z, \tag{8.172}$$

$$D_{lm} = \frac{1}{Z}\sum_j\left(X_{lj}R_{jm}e^{-\beta E_m} - R_{lj}e^{-\beta E_j}X_{jm} - X_{lj}e^{-\beta E_j}(R^\dagger)_{jm} + e^{-\beta E_l}(R^\dagger)_{lj}X_{jm}\right). \tag{8.173}$$

Because of $X_{ml} = X_{lm}^*$, $(R^\dagger)_{lm} = \langle m|R|l\rangle^*$. Then, using the KMS relation

$$\Psi(\omega) = e^{-\beta\hbar\omega}\Psi(-\omega), \tag{8.174}$$

we have

$$(R^\dagger)_{lm} = e^{\beta(E_l-E_m)}R_{lm}. \tag{8.175}$$

Substituting the relation, we have $D_{lm} = 0$. Thus, the canonical distribution is the stationary state.

8.4.3 Redfield Type for a Single Spin

We adopt the coupling to the bath in the form $X(b+b^\dagger)$ with

$$X = a_x S_x + a_y S_y + a_z S_z = \frac{1}{2}\begin{pmatrix} a_z & a_x - ia_y \\ a_x + ia_y & -a_z \end{pmatrix}. \tag{8.176}$$

The eigenvalues and eigenvector of the Hamiltonian are

$$\mathcal{H}\begin{pmatrix} x_\pm \\ y_\pm \end{pmatrix} = \pm\frac{1}{2}\sqrt{B_z^2 + B_x^2 + B_y^2}\begin{pmatrix} x_\pm \\ y_\pm \end{pmatrix} \equiv \pm E\begin{pmatrix} x_\pm \\ y_\pm \end{pmatrix}. \tag{8.177}$$

With the unitary transformation:

$$U = \begin{pmatrix} x_+ & x_- \\ y_+ & y_- \end{pmatrix}, \quad U^{-1} = \begin{pmatrix} x_+ & y_+ \\ x_- & y_- \end{pmatrix}, \tag{8.178}$$

the Hamiltonian is diagonalized as

$$U^{-1}\mathcal{H}U = \frac{1}{2}\begin{pmatrix} E & 0 \\ 0 & -E \end{pmatrix}. \tag{8.179}$$

For the dissipation term

$$D[\rho] = -\frac{\pi\lambda^2}{\hbar}\left(XR\rho - R\rho X + \rho R^\dagger X - X\rho R^\dagger\right), \tag{8.180}$$

we now obtain the matrix R. The operator X is expressed on the basis of eigenstates of $\mathcal{H}$ as

$$U^{-1}XU \equiv \begin{pmatrix} X_{11} & X_{12} \\ X_{21} & X_{22} \end{pmatrix}. \tag{8.181}$$

Defining

$$\Delta E = 2E, \tag{8.182}$$

the matrix R is given by

$$U^{-1}RU = \frac{1}{2}\begin{pmatrix} 0 & I_0(\Delta E)^\alpha \frac{1}{e^{\beta\Delta E}-1}X_{12} \\ I_0(\Delta E)^\alpha \frac{e^{\beta\Delta E}}{e^{\beta\Delta E}-1}X_{21} & 0 \end{pmatrix}, \tag{8.183}$$

i.e.,

$$R = \frac{1}{2}U\begin{pmatrix} 0 & I_0(\Delta E)^\alpha \frac{1}{e^{\beta\Delta E}-1}X_{12} \\ I_0(\Delta E)^\alpha \frac{e^{\beta\Delta E}}{e^{\beta\Delta E}-1}X_{21} & 0 \end{pmatrix}U^{-1} \equiv \begin{pmatrix} R_{11} & R_{12} \\ R_{21} & R_{22} \end{pmatrix}. \tag{8.184}$$

where we assume $\alpha > 0$ and $I(\omega = 0) = 0$.

Using the explicit forms of $\mathcal{H}$ U,X,R, $D[\rho]$ (8.180) is given by a 2×2 matrix of the form:

$$D[\rho] = \begin{pmatrix} A & B - iC \\ B + iC & -A \end{pmatrix}, \tag{8.185}$$

which has the same form as that of Lindblad ones (8.144). In this formula, the stationary state is guaranteed to be the equilibrium state at a given temperature, which was explained in Sect. 8.4.2.1.

8.4.4 Bloch Equation

To study dissipation effects on a single spin, the Bloch equation is often used. We now study a relationship between the above obtained equations (8.139) and (8.185) for single spin systems and the Bloch equation for the magnetizations:

$$\dot{M}_x = B_y M_z - B_z M_y - \frac{1}{T_2} M_x, \tag{8.186}$$

$$\dot{M}_y = B_z M_x - B_x M_z - \frac{1}{T_2} M_y, \tag{8.187}$$

$$\dot{M}_z = B_x M_y - B_y M_x - \frac{1}{T_1}(M_z - M_z^{(\mathrm{eq})}), \tag{8.188}$$

where

$$M_x = \mathrm{Tr} S_x \rho = \frac{\rho_{12} + \rho_{21}}{2}, \quad M_y = i\frac{\rho_{12} - \rho_{21}}{2}, \quad M_z = \frac{\rho_{11} - \rho_{22}}{2} = 2\rho_{11} - 1. \tag{8.189}$$

Now, let the Hamiltonian be

$$\mathcal{H} = \begin{pmatrix} a & b \\ b^* & -a \end{pmatrix}, \tag{8.190}$$

where

$$a = \frac{1}{2} B_z, \quad b = \frac{1}{2}(B_x - iB_y). \tag{8.191}$$

Then, the part of the quantum motion is expressed as

$$[\mathcal{H}, \rho] = \begin{pmatrix} a & b \\ b^* & -a \end{pmatrix} \begin{pmatrix} \rho_{11} & \rho_{12} \\ \rho_{21} & \rho_{22} \end{pmatrix} - \begin{pmatrix} \rho_{11} & \rho_{12} \\ \rho_{21} & \rho_{22} \end{pmatrix} \begin{pmatrix} a & b \\ b^* & -a \end{pmatrix}$$

$$= \begin{pmatrix} a\rho_{11} + b\rho_{21} & a\rho_{12} + b\rho_{22} \\ b^*\rho_{11} - a\rho_{21} & b^*\rho_{12} - a\rho_{22} \end{pmatrix} - \begin{pmatrix} a\rho_{11} + b^*\rho_{21} & b\rho_{11} - a\rho_{12} \\ a\rho_{21} + b^*\rho_{22} & b\rho_{21} - a\rho_{22} \end{pmatrix}$$

$$= \begin{pmatrix} b\rho_{21} - b^*\rho_{12} & 2a\rho_{12} + b(\rho_{22} - \rho_{11}) \\ -2a\rho_{21} + b^*(\rho_{11} - \rho_{22}) & b^*\rho_{12} - b\rho_{21} \end{pmatrix}, \tag{8.192}$$

and the dissipation part of the Lindblad equation (8.144) is given by

$$D[\rho] = -\begin{pmatrix} 2\alpha_-\rho_{11} - 2\alpha_+\rho_{22} & (\alpha_+ + \alpha_- + \alpha_z)\rho_{12} \\ (\alpha_+ + \alpha_- + \alpha_z)\rho_{21} & 2\alpha_+\rho_{22} - 2\alpha_-\rho_{11} \end{pmatrix}. \tag{8.193}$$

Thus, the equations of motion of the matrix elements are

$$\dot{\rho}_{11} = -i(b\rho_{21} - b^*\rho_{12}) - (2\alpha_-\rho_{11} - 2\alpha_+\rho_{22}) \tag{8.194}$$

$$\dot{\rho}_{12} = -i(2a\rho_{12} + b(\rho_{22} - \rho_{11})) - (\alpha_+ + \alpha_- + \alpha_z)\rho_{12} \tag{8.195}$$

$$\dot{\rho}_{21} = i(2a\rho_{21} + b^*(\rho_{22} - \rho_{11})) - (\alpha_+ + \alpha_- + \alpha_z)\rho_{21} \tag{8.196}$$

and

$$\dot{\rho}_{22} = -\dot{\rho}_{11}. \tag{8.197}$$

And similar relations hold for the Redfield equation (8.185).

These equations (8.194)–(8.196) give the Bloch equation (8.186)–(8.188) with the relations (8.189), where

$$\frac{1}{T_2} = (\alpha_+ + \alpha_- + \alpha_z), \quad \frac{1}{T_2} = 2(\alpha_+ + \alpha_-), \quad M_z^{(\mathrm{eq})} = \frac{\alpha_+ - \alpha_-}{\alpha_+ + \alpha_-}. \tag{8.198}$$

Thus, for the single spin case, all the three methods are equivalent.

8.4.5 Under a Time-Dependent Field

The above-mentioned methods use the dissipation term ($D[\rho]$) which brings the system to the equilibrium state (the canonical distribution) for the given Hamiltonian. When the external field changes in time, the Hamiltonian changes in time. In this case, how to modify the form of $D[\rho]$ is not trivial. The simplest way is that we use the form $D[\rho(t), t]$ for the instantaneous Hamiltonian $\mathcal{H}(t)$. For the validity of this choice, we assume that the change of the Hamiltonian is much slow than the time scale of motions of the dissipative mechanism. The time scale is a really subtle point, and we need to be careful. But in practice, what we can do is only in this case.

8.5 Path-Integral Method

We give a brief review of the path-integral method [193, 194], and refer to relations to the partition function and time-evolution operator in a stochastic process [195].

8.5.1 One Particle Problem

First, we consider a system with a particle moving in a potential function $V(x)$. The Hamiltonian is given by

$$\mathcal{H} = \frac{p^2}{2m} + V(x), \tag{8.199}$$

where x and p are the position and the momentum, respectively. The Schrödinger equation is given by

$$i\hbar \frac{\partial \psi(x,t)}{\partial t} = -\frac{\hbar^2}{2m}\frac{\partial^2 \psi(x,t)}{\partial x^2} + V(x)\psi(x,t). \tag{8.200}$$

The solution is formally given by

$$\psi(x,t) = U(t,0)\psi(x,0), \quad U(t,0) = \mathrm{T}e^{-i\mathcal{H}t/\hbar}\psi(x,0), \tag{8.201}$$

where T denotes the time-ordered exponential. We may write the time-evolution function $U(t, 0)$ as

$$\mathrm{T}e^{-i\mathcal{H}t/\hbar} = \left(e^{-i\mathcal{H}\Delta t/\hbar}\right)^M, \quad M = \frac{t}{\Delta t}. \tag{8.202}$$

Substituting the complete basis sets $\int dx|x\rangle\langle x|$ between the exponentials, the matrix element $U(t,0)_{x_M,x_0}$ is given by

$$\langle x_M|e^{-i\mathcal{H}\Delta t/\hbar} \int dx_{M-1}|x_{M-1}\rangle\langle x_{M-1}|\langle x_{M-1}|e^{-i\mathcal{H}\Delta t/\hbar} \int dx_{M-2}|x_{M-2}\rangle\langle_{M-2}| \cdots$$

$$\cdots |x_1\rangle\langle x_1|e^{-i\mathcal{H}\Delta t/\hbar}|x_0\rangle. \tag{8.203}$$

Using the complete set of eigenstates for the momentum,

$$\langle x_{k+1}|e^{-i\mathcal{H}\Delta t/\hbar}|x_k\rangle = \langle x_{k+1}| \int dp_k|p_k\rangle\langle p_k|e^{-i\mathcal{H}\Delta t/\hbar}|x_k\rangle. \tag{8.204}$$

For small interval Δt, we may write

$$e^{-i\mathcal{H}\Delta t} = 1 - \frac{i}{\hbar}\mathcal{H} \tag{8.205}$$

and then, (8.204) is given by

$$\langle x_{k+1}|e^{-i\mathcal{H}\Delta t/\hbar}|x_k\rangle = \int dp_k \langle x_{k+1}|p_k\rangle\langle p_k|x_k\rangle \left(1 - \frac{i}{\hbar}\left(\frac{p_k^2}{2m} + V(x_k)\right)\right). \tag{8.206}$$

Using the relation:

$$\langle x|p\rangle = \sqrt{\frac{1}{2\pi\hbar}}e^{ipx/\hbar}, \tag{8.207}$$

the matrix element is given by

$$\langle x_{k+1}|p_k\rangle\langle p_k|x_k\rangle = \frac{1}{2\pi\hbar}e^{ip_k(x_{p+1}-x_k)/\hbar}. \tag{8.208}$$

Thus, the time-evolution operator (8.201) is given in the form

$$U(t,0)_{x_M,x_0} = \int dx_{M-1}\cdots\int dx_1 \int dp_{M-1}\cdots\int dp_1$$

$$\exp\left(\frac{i}{\hbar}\sum_{k=1}^{M-1}\left(p_k(x_{p+1}-x_k) - \Delta t\left(\frac{p_k^2}{2m} + V(x_k)\right)\right)\right). \tag{8.209}$$

Using the following notations,

$$\int dx_{M-1}\cdots\int dx_1 = \int \mathcal{D}(x), \quad \int dp_{M-1}\cdots\int dp_1 = \int \mathcal{D}(p), \tag{8.210}$$

and

$$x_{k+1} - x_k = \frac{x_{k+1}-x_k}{\Delta t}\Delta t = \dot{x}, \tag{8.211}$$

we have the expression:

$$U(t,0)_{x_M,x_0} = \int \mathcal{D}(x)\int \mathcal{D}(p)\exp\left(\frac{i}{\hbar}\int_0^t dt\left(p(t')\dot{x}(t') - \mathcal{H}(p(t'),x(t'))\right)\right). \tag{8.212}$$

The integration gives the sum of all the paths of x and p. For example, a path of the position changes as

$$x_0 \to x_1 \to x_2\cdots \to x_{M-1} \to x_M. \tag{8.213}$$

Using the notation of the action $S(p,x)$ in the analytical mechanics, it is expressed as

$$U(t,0)_{x_M,x_0} = \int \mathcal{D}(x) \int \mathcal{D}(p) \exp\left(\frac{i}{\hbar} S(p,x))\right). \tag{8.214}$$

If we integrate over $p(t')$, we have the form in the variable of position which is often used:

$$U(t,0)_{x_M,x_0} = \int \mathcal{D}(x) \exp\left(\frac{i}{\hbar}\int_0^t dt \left(\frac{m}{2}\dot{x}(t')^2 - V(x')\right)\right)$$

$$= \int \mathcal{D}(x) \exp\left(\frac{i}{\hbar}\int_0^t dt L(\dot{x},x)\right), \tag{8.215}$$

where $L(\dot{x},x)$ is regarded as the Lagrangian.

8.5.2 *Partition Function at a Finite Temperature*

In quantum mechanical systems, the partition function is expressed as

$$Z = \mathrm{Tr}e^{-\beta\mathcal{H}} = \sum_{\text{allstates } S} \langle S|e^{-\beta\mathcal{H}}|S\rangle, \tag{8.216}$$

where $\{|S\rangle\}$ is a normalized orthogonal basis set of the system. If we take the eigenstates $\{|\psi_n\rangle\}$ of the Hamiltonian as the basis set, the partition function is given by the standard form

$$Z = \sum_{\text{states}\psi_n} \langle\psi_n|e^{-\beta\mathcal{H}}|\psi_n\rangle = \sum_n e^{-\beta E_n}, \tag{8.217}$$

where E_n is the nth eigenvalue. If we take another basis set which is not the set of eigenstates of the system, (8.216) is still valid, but it is not simple to calculate the value of

$$Z(S) \equiv \langle S|e^{-\beta\mathcal{H}}|S\rangle. \tag{8.218}$$

Using a trivial relation

$$e^{-\beta\mathcal{H}} = \left(e^{-\frac{\beta}{n}\mathcal{H}}\right)^n, \tag{8.219}$$

and inserting an identity

$$I = \sum_k |S_k\rangle\langle S_k| \tag{8.220}$$

between each $e^{-\frac{\beta}{n}\mathcal{H}}$, we have

$$Z(S)=\sum_{k_1}\sum_{k_2}\cdots\sum_{k_{n-1}}\langle S|e^{-\frac{\beta}{n}\mathcal{H}}|S_{k_1}\rangle\langle S_{k_1}|e^{-\frac{\beta}{n}\mathcal{H}}\cdots|S_{k_{n-1}}\rangle\langle S_{k_{n-1}}|e^{-\frac{\beta}{n}\mathcal{H}}|S\rangle, \quad (8.221)$$

which has the same structure of the transfer matrix. The sums over $\{S_{k_1}, S_{k_2}, \ldots, S_{k_{n-1}}\}$ correspond to the sum over all the possible configurations fixing the initial and final states to be $|S\rangle$.

We divide the Hamiltonian as

$$\mathcal{H}=\mathcal{H}_{\text{diagonal}}+\mathcal{H}_{\text{off-diagonal}}, \quad (8.222)$$

where $\mathcal{H}_{\text{diagonal}}$ and $\mathcal{H}_{\text{off-diagonal}}$ are diagonal and off-diagonal parts in the basis set $\{|S\rangle\}$. Then, the Boltzmann factor is approximated as

$$e^{-\frac{\beta}{n}\mathcal{H}}\simeq e^{-\frac{\beta}{n}\mathcal{H}_{\text{diagonal}}}e^{-\frac{\beta}{n}\mathcal{H}_{\text{off-diagonal}}} \quad (8.223)$$

for small β/n. Then, the matrix element is given by

$$\langle S|e^{-\frac{\beta}{n}\mathcal{H}}|S'\rangle=e^{-\frac{\beta}{n}\langle S|\mathcal{H}_{\text{diagonal}}|S\rangle}\langle S|e^{-\frac{\beta}{n}\mathcal{H}_{\text{off-diagonal}}}|S'\rangle. \quad (8.224)$$

The term $\langle S|\mathcal{H}_{\text{diagonal}}|S\rangle\delta_{S,S'}$ gives the contribution from the kth layer, and the term $\langle S|e^{-\frac{\beta}{n}\mathcal{H}_{\text{off-diagonal}}}|S'\rangle$ gives the contribution from the connection part of the kth and $(k+1)$th layers. This structure reminds us of the transfer matrix of the $(d+1)$-dimensional lattice which consists of the original d-dimensions of the system and a new direction of n which is called the imaginary time axis. This idea was used for the quantum Monte Carlo simulation [118] as explained in Sect. 6.12. This $(d+1)$-dimensional expression is a path-integral representation of the partition function.

When we express the transfer matrix in the above argument by $\hat{T}$, the partition function is given by

$$Z=\text{Tr}\hat{T}^n=\sum_k\lambda_k^n\simeq\lambda_1^n. \quad (8.225)$$

Thus, the free energy par spin is

$$f(\beta)=-k_{\text{B}}Tn\ln\lambda_1. \quad (8.226)$$

Because, for large n,

$$e^{-\frac{\beta}{n}\mathcal{H}}\simeq 1-\frac{\beta\mathcal{H}}{n}, \quad \text{and} \quad \lambda_1\propto\frac{1}{n}, \quad (8.227)$$

and thus $f(\beta)$ is independent of n, and it agrees with the one obtained from the original form of the partition function (8.216).

For a single-particle system, the partition function is given by

$$Z = \int \mathcal{D}(x(u)) \exp\left(-\frac{1}{\hbar}\int_0^\beta \left[\frac{m\dot{x}(u)^2}{2} + V(x(u))du\right]\right). \tag{8.228}$$

A path which gives the largest weight is called the most probable path. As the first approximation, the functional integral over $\mathcal{D}(x(u))$ is given by the contribution of the most probable path. For a better approximation, one may include contributions from harmonic fluctuation around the most probable path.

8.5.3 *Onsager–Machlup Formula for Stochastic Process*

L. Onsager gave the famous Onsager's reciprocal theory [196] as the relation among kinetic coefficients of linear response equations of fluxes (i.e., currents) given as functions of the external fields. The fluxes are given by $\{\dot{\alpha}_i\}$, i.e., the time derivatives of quantities $\{\alpha_i\}$. There, the excess entropy due to $\{\alpha_i\}$ is given by a quadratic form:

$$S(\alpha_1, \cdots, \alpha_n) = -\frac{1}{2}\sum_{i,j} s_{ij}\alpha_i\alpha_j, \quad \alpha_{ij} = \alpha_{ji} \tag{8.229}$$

and the thermodynamic force is defined by

$$X_i = \frac{\partial S}{\partial \alpha_i} = -\sum_i s_{ij}\alpha_j. \tag{8.230}$$

Postulating that the dynamics of microscopic fluctuation follows the macroscopic relation form given by

$$\sum_j R_{ij}\dot{\alpha}_j = X_i \quad \alpha_i = \sum_j L_{ij}X_j, \quad R = L^{-1}, \tag{8.231}$$

we have the relation:

$$X_i = \sum_j R_{ij}\dot{\alpha}_j = -\sum_i s_{ij}\alpha_j, \tag{8.232}$$

which is a generalized form of the Debye (exponential decay) model.

To study the irreversible process, Onsager and Machlup introduced a stochastic process given by

$$\sum_j R_{ij}\dot{\alpha}_j = -\sum_i s_{ij}\alpha_j + \epsilon(t), \tag{8.233}$$

where $\epsilon(t)$ is a random force and studied the path of the α's in time under the influence of the random force. With some considerations, they adopted the random force to be the white-Gaussian noise (8.80). Then they considered a cumulative distribution function of values in a time-sequence $\{\boldsymbol{\alpha}\}$ at $\{t_k\}$, $k = 0, 2, \ldots, p$ along the stochastic process:

$$F_p\left(\boldsymbol{\alpha}(t_0), \boldsymbol{\alpha}(t_1), \ldots, \boldsymbol{\alpha}(t_p)\right). \tag{8.234}$$

For simplicity, let us consider the single α case:

$$R\dot{\alpha} = -s\alpha + \xi(t). \tag{8.235}$$

This is the same form as the Ornstein–Uhlenbeck model (8.84):

$$m\frac{dv}{dt} = -kv + \xi(t),$$

by identifying

$$R = m, \quad s = k. \tag{8.236}$$

Because the propagator was given by (8.96)

$$P(v(\Delta t), v(0)) \simeq \sqrt{\frac{m^2}{4\pi D\Delta t}} \exp\left(-\frac{m^2}{4}\frac{(v(t) - \langle v(\Delta t)\rangle)^2}{D\Delta t}\right),$$

the cumulative distribution function is given by

$$F_p\left(\boldsymbol{\alpha}(t_0), \boldsymbol{\alpha}(t_1), \ldots, \boldsymbol{\alpha}(t_p)\right) = P(\alpha(t_p), \alpha(t_{p-1})) \times \cdots \times P(\alpha(t_1), \alpha(t_0)). \tag{8.237}$$

Noting that

$$\frac{(\alpha(\Delta t_p) - \langle\alpha(\Delta t_p)\rangle)^2}{\Delta t} = \left(\frac{\alpha(\Delta t_p) - \alpha(\Delta t_{p-1})e^{-\gamma\Delta t}}{\Delta t}\right)^2 \Delta t$$

$$\simeq \left(\frac{\alpha(\Delta t_p) - \alpha(\Delta t_{p-1})(1 - \gamma\Delta t)}{\Delta t}\right)^2 \Delta t = (\dot{\alpha}(t) + \gamma\alpha(t))^2\, \Delta t, \tag{8.238}$$

where $\gamma = s/R$, the cumulative distribution function is given by

$$F_p\left(\boldsymbol{\alpha}(t_0), \boldsymbol{\alpha}(t_1), \ldots, \boldsymbol{\alpha}(t_p),\right) = \left(\sqrt{\frac{R^2}{4\pi D\Delta t}}\right)^p \times$$

$$\exp\left(-\frac{R^2}{4D}[\dot{\alpha}(t_p) + \gamma\alpha(t_{p-1})]^2\Delta t\right) \cdots \exp\left(-\frac{R^2}{4D}[\dot{\alpha}(t_1) + \gamma\alpha(t_0)]^2\Delta t\right). \tag{8.239}$$

Taking the limit $\Delta t \to 0$, we have the cumulative distribution function as a functional of the function of $\alpha(t)$, $t_p \geq t \geq t_0$:

$$F_p\left[\boldsymbol{\alpha}(t)\right] \propto \exp\left(-\frac{R^2}{4D}\int_{t_0}^{t_p}[\dot{\alpha}(t)+\gamma\alpha(t)]^2 dt\right). \tag{8.240}$$

This is a kind of path-integral formulation with the Gaussian path-integral form:

$$F_p\left[\boldsymbol{\alpha}(t)\right] \propto \exp\left(-\int_{t_0}^{t_p}\mathcal{L}(\boldsymbol{\alpha}(t), \dot{\boldsymbol{\alpha}}(t))dt\right). \tag{8.241}$$

The transition probability from $\alpha(t_0)$ at $t = t_0$ to $\alpha(t_p)$ at $t = t_p$ is given by summing up all the possible paths in the period $t_p \geq t \geq t_0$, which is given by the solution of the minimum value of the integral.

When there are many variables, the relation is still a linear equation. Then we use a representation which diagonalizes the set of equations and apply the present analysis, then we have a similar relation with the variables. The part of Lagrangian $\mathcal{L}$ of the weight of the path in the stochastic process is called the 'Onsager–Machlup function' [195].

8.6 WKB Approximation

Tunneling phenomena through an energy barrier (Fig. 8.6) are estimated by the the Wentzel–Kramers–Brillouin (WKB) approximation [60, 197].

8.6.1 *Semiclassical Approximation*

First, we briefly review the WKB approximation. We consider the Schrödinger equation for a particle with mass m in a potential $V(x)$ in one dimension:

$$i\hbar\frac{\partial}{\partial t}\psi(x,t) = -\frac{\hbar^2}{2m}\frac{\partial^2\phi(x)}{\partial x^2}\psi(x,t) + V(x)\psi(x,t). \tag{8.242}$$

Let the wave function be in the form:

$$\psi(x,t) = \exp\left(\frac{i}{\hbar}W(x,t)\right). \tag{8.243}$$

Substituting the form (8.243) to the Schrödinger equation (8.242), we have

$$-\frac{\partial}{\partial t}W(X,t)=\left(\frac{1}{2m}\frac{\partial W(x,t)}{\partial x}\right)^2+V(x)-\frac{i\hbar}{2m}\frac{\partial^2 W(x,t)(x)}{\partial x^2}. \tag{8.244}$$

In the limit $\hbar \to 0$, we have the Hamilton–Jacobi equation

$$-\frac{\partial}{\partial t}W_c(X,t)=\left(\frac{1}{2m}\frac{\partial W_c(x,t)}{\partial x}\right)^2+V(x), \tag{8.245}$$

and $W_c(x,t)$ is Hamilton's principal function, in other words, the action in the notation of the analytical mechanics. When the total energy E is conserved, Hamilton's principal function is given in the form

$$W_c(x,t)=S_c(x)-Et, \tag{8.246}$$

where $S_c(x)$ is Hamiltonian's characteristic function which is also called action.

The time-independent Schrödinger equation is

$$\frac{d^2\phi(x)}{dx^2}+\frac{2m}{\hbar^2}(E-V(x))\phi(x)=0. \tag{8.247}$$

Expressing the wave function in the form:

$$\phi(x)=\exp\left(\frac{i}{\hbar}S(x)\right), \tag{8.248}$$

we have the equation for $S(x)$

$$\left(\frac{dS(x)}{dx}\right)^2-i\hbar\frac{d^2S(x)}{dx^2}-2m(E-V(x))=0. \tag{8.249}$$

Here, the term $i\hbar\frac{d^2S(x)}{dx^2}$ is a quantum correction to the classical relation for Hamiltonian's characteristic function which is given by

$$\frac{dS_c(x)}{dx}=p=\pm\sqrt{2m(E-V(x))}. \tag{8.250}$$

Now, we expand $S(x)$ with respect to $\hbar$:

$$S(x)=S_c+\frac{\hbar}{i}S_1+\left(\frac{\hbar}{i}\right)^2S_2+\cdots, \tag{8.251}$$

and substitute it to (8.279). In the order $O(\hbar^0)$, we have (8.250), and in the order $O(\hbar)$

$$\frac{dS_1}{dx} = -\frac{1}{2}\frac{\frac{d^2S_c(x)}{dx^2}}{\frac{dS_c(x)}{dx}} = -\frac{1}{2}\frac{d}{dx}\ln\left(\frac{dS_c(x)}{dx}\right). \tag{8.252}$$

From these, we have

$$S_1(x) = -\frac{1}{2}\log p(x) + \text{constant}. \tag{8.253}$$

Thus, up to the first order of $\hbar$,

$$S(x) = \pm\int p(x')dx' - \frac{\hbar}{i}\frac{1}{2}\log p(x), \tag{8.254}$$

and the wave function is given in the form with coefficients A and B:

$$\phi(x) = \frac{A}{\sqrt{p(x)}}\exp\left[\frac{i}{\hbar}\int^x p(x)dx\right] + \frac{B}{\sqrt{p(x)}}\exp\left[-\frac{i}{\hbar}\int^x p(x)dx\right]. \tag{8.255}$$

In the case $E < V(x)$, classically no motion is allowed. However, in quantum systems the wave function exists, and setting

$$\rho(x) = \sqrt{2m(V(x) - E)}, \tag{8.256}$$

the wave function is given by

$$\phi(x) = \frac{A'}{\sqrt{\rho(x)}}\exp\left[-\frac{1}{\hbar}\int^x \rho(x)dx\right] + \frac{B'}{\sqrt{\rho(x)}}\exp\left[\frac{1}{\hbar}\int_x \rho(x)dx\right]. \tag{8.257}$$

This is an approximation for small $\hbar$, and it is called the semiclassical approach.

8.6.2 Connection Formula

The wave functions (8.255) and (8.257) must be connected properly for both sides of $E - V(x) > 0$ and $E - V(x) < 0$ (Fig. 8.5). The point ($x = a$) at which $|E - V(x)| = 0$ is called the 'turning point'. Near the point, $S_1(x)$ becomes large and the approximation is not good. But, apart from the region near the turning point, we can choose the coefficients to connect these wave functions properly. To find such coefficients, we obtain a correct wave function around the turning point. Near the turning point ($x = a$), the potential $V(x)$ can be linearized and has the form with a coefficient F:

$$V(x) = E \pm F(x - a), \quad F > 0, \tag{8.258}$$

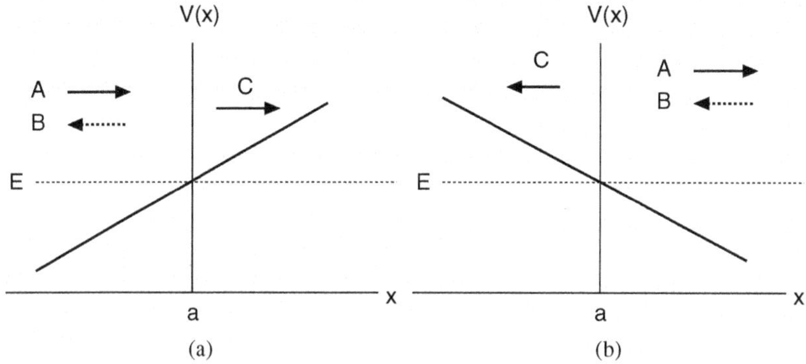

Fig. 8.5 The potentials: **a** $V(x) = E + F(x - a)$, and **b** $V(x) = E - F(x - a)$

as depicted in Fig. 8.5a and b, respectively. For this potential, the Schrödinger equation is given by

$$-\frac{\hbar^2}{2m}\frac{d^2}{dx^2}\phi(x) + Fx\phi(x) = E\phi(x), \tag{8.259}$$

and its solution is given by the Airy function.

By making use of asymptotic forms of the Airy function, we have the coefficients for properly corrected wave functions.

8.6.2.1 Connection in Fig. 8.5a

If the wave function in the right of the point a in Fig. 8.5a is given by

$$\psi_1(x > a) = \frac{1}{\sqrt{\rho(x)}} \exp\left[-\frac{1}{\hbar}\int_a^x \rho(x')dx'\right], \tag{8.260}$$

it is connected to

$$\psi_1(x < a) = \frac{2}{\sqrt{p(x)}} \sin\left[\frac{1}{\hbar}\int_x^a p(x')dx' + \frac{\pi}{4}\right]. \tag{8.261}$$

This relation is called the 'connection formula' for one turning point. By making use of the fact that the Wronskian

$$W(x) = \begin{vmatrix} \psi_1(x), \psi_2(x) \\ \psi_1'(x), \psi_2'(x) \end{vmatrix} \tag{8.262}$$

is a constant (independent of x) when $\psi_1(x)$ and $\psi_2(x)$ are independent, the wave function independent of $\psi_1(x)$ is given by

$$\psi_2(x > a) = \frac{1}{\sqrt{\rho(x)}} \exp\left[\frac{1}{\hbar}\int_a^x \rho(x')dx'\right], \tag{8.263}$$

$$\psi_2(x < a) = \frac{1}{\sqrt{p(x)}} \cos\left[\frac{1}{\hbar}\int_x^a p(x')dx' + \frac{\pi}{4}\right], \tag{8.264}$$

where it should be noted that factor 2 in (8.261) disappears. From these relations (8.261) and (8.264), the wave function

$$\frac{1}{\sqrt{p(x)}} \exp\left[\pm\frac{i}{\hbar}\int_x^a p(x')dx' \pm i\frac{\pi}{4}\right], \quad x < a \tag{8.265}$$

is connected to

$$\frac{1}{\sqrt{\rho(x)}} \exp\left[\frac{1}{\hbar}\int_a^x \rho(x')dx'\right] \pm \frac{i}{2\sqrt{\rho(x)}} \exp\left[-\frac{1}{\hbar}\int_a^x \rho(x')dx'\right]. \tag{8.266}$$

These relations are useful to obtain the coefficients to connect the wave functions on both sides.

8.6.2.2 Connection in Fig. 8.5b

In Fig. 8.5b, we have similar connection formula. The wave function on the left of the point:

$$\psi_1(x < a) = \frac{1}{\sqrt{\rho(x)}} \exp\left[-\frac{1}{\hbar}\int_x^a \rho(x')dx'\right] \tag{8.267}$$

is connected to

$$\psi_1(x > a) = \frac{2}{\sqrt{p(x)}} \sin\left[\frac{1}{\hbar}\int_a^x p(x')dx' + \frac{\pi}{4}\right]. \tag{8.268}$$

The independent solution is

$$\psi_2(x < a) = \frac{1}{\sqrt{\rho(x)}} \exp\left[\frac{1}{\hbar}\int_x^a \rho(x')dx'\right], \tag{8.269}$$

$$\psi_2(x > a) = \frac{1}{\sqrt{p(x)}} \cos\left[\frac{1}{\hbar}\int_a^x p(x')dx' + \frac{\pi}{4}\right]. \tag{8.270}$$

From these,

$$\frac{1}{\sqrt{p(x)}} \exp\left[\pm\frac{i}{\hbar}\int_a^x p(x')dx' \mp i\frac{\pi}{4}\right] \equiv \frac{1}{\sqrt{p(x)}} \exp\left[\pm iZ\right], \quad x < a, \tag{8.271}$$

where

$$Z = \frac{1}{\hbar}\int_a^x p(x')dx' - \frac{\pi}{4} \tag{8.272}$$

is connected to[1]

$$\frac{1}{2\sqrt{\rho(x)}} \exp\left[-\frac{1}{\hbar}\int_x^a \rho(x')dx'\right] \mp \frac{i}{\sqrt{\rho(x)}} \exp\left[\frac{1}{\hbar}\int_x^a \rho(x')dx'\right], \quad x > a. \tag{8.273}$$

8.6.3 Bound State

When the state is bounded in a potential well, the wave function must decay in the region $E < V(x)$. Let $x = a$ and b $(a < b)$ be the turning points where $E = V(x)$. There, we need to connect the wave function at $x = a$ and b. Because of (8.267) and (8.260), the wave function between a and b is given by (8.261) and also by (8.268):

1

$$\sin(Z) = \frac{e^{iZ} - e^{-iZ}}{2i} \to -\frac{\sqrt{p}}{\sqrt{\rho}} e^{\frac{1}{\hbar}\int_x^a \rho(x')dx'}, \quad \cos(Z) = \frac{e^{iZ} + e^{-iZ}}{2} \to \frac{\sqrt{p}}{2\sqrt{\rho}} e^{-\frac{1}{\hbar}\int_x^a \rho(x')dx'}$$

$$\frac{1}{\sqrt{p}} e^{\pm iZ} \to \frac{1}{2\sqrt{\rho}} e^{-\frac{1}{\hbar}\int_x^a \rho(x')dx'} \mp i \frac{1}{\sqrt{\rho}} e^{\frac{1}{\hbar}\int_x^a \rho(x')dx'}.$$

$$\sin\left[\frac{1}{\hbar}\int_x^b p(x')dx' + \frac{\pi}{4}\right] = \sin\left[\frac{1}{\hbar}\int_a^b p(x')dx' - \frac{1}{\hbar}\int_a^x p(x')dx' + \frac{\pi}{4}\right]$$

$$= \sin\left[\frac{1}{\hbar}\int_a^x p(x')dx' - \frac{1}{\hbar}\int_a^b p(x')dx' - \frac{\pi}{4} + \pi\right]$$

$$= -\sin\left[-\frac{1}{\hbar}\int_a^b p(x')dx' + \frac{1}{\hbar}\int_a^x p(x')dx' - \frac{\pi}{4}\right]. \tag{8.274}$$

This requires that

$$-\frac{1}{\hbar}\int_a^b p(x')dx' + \pi = 2\pi n, \quad n : \text{integer}, \tag{8.275}$$

i.e.,

$$\int_a^b p(x')dx' = 2\pi\hbar\left(n + \frac{1}{2}\right), \tag{8.276}$$

which agrees with the Bohr–Sommerfeld quantization rule.

If we try to obtain a WKB connection for the energy which does not satisfy this condition, we do not have a solution which decays outside. This is consistent with the way to obtain the bound state in solving the Schrödinger equation.

If we consider the case of metastable potential, this condition is not satisfied because there is a component of exponentially increasing wave function at $x > b$.

8.6.4 Transmission Coefficient by WKB Approximation

We study transition probability through a potential barrier as schematically depicted in Fig. 8.6. Using the connection formula, we obtain the transmission coefficient $T = |C|^2/|A|^2$ over the potential barrier.

We consider the case where the output wave function in $x > b$ is

$$\phi_{\text{out}}(x) = \frac{2C}{\sqrt{p(x)}} \exp\left[\frac{i}{\hbar}\int_b^x p(x')dx' - i\frac{\pi}{4}\right], \tag{8.277}$$

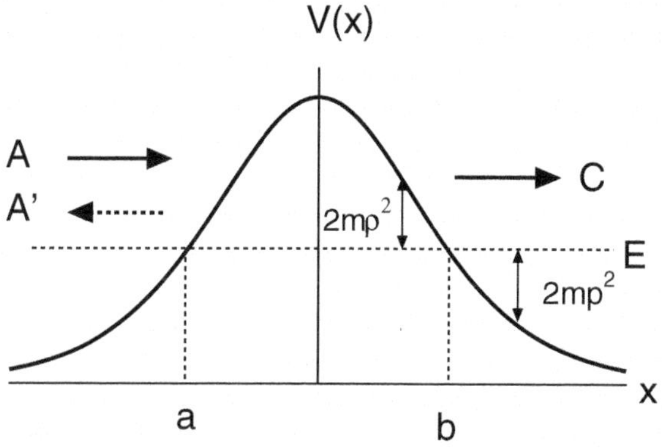

Fig. 8.6 Tunneling through an energy barrier $V(x)$. Here, $\psi(x > b) = \frac{C}{\sqrt{p(x)}} \exp\left[\frac{i}{\hbar}\int_b^x p(x')dx' - i\frac{\pi}{4}\right]$ and $\phi(x < a) = \frac{A}{\sqrt{p}} \exp\left[\frac{i}{\hbar}\int_a^x p(x')dx' - \frac{\pi}{4}\right] + \frac{A'}{\sqrt{p}} \exp\left[-\frac{i}{\hbar}\int_a^x p(x')dx' + \frac{\pi}{4}\right]$

where we put the phase for the convenience of calculation. By (8.273), this wave function is connected to

$$\phi_{\text{barrier}}(x) = \frac{C}{\sqrt{\rho(x)}}\left[\exp\left(-\frac{1}{\hbar}\int_x^b \rho(x')dx'\right) - 2i\exp\left(\frac{1}{\hbar}\int_x^b \rho(x')dx'\right)\right]. \tag{8.278}$$

Using the following notation

$$S = \frac{1}{\hbar}\int_a^b \rho(x')dx', \tag{8.279}$$

we may rewrite it as[2]

$$\phi_{\text{barrier}}(x) = \frac{C}{\sqrt{\rho(x)}}\left[e^{-S}\exp\left(\frac{1}{\hbar}\int_a^x \rho(x')dx'\right) - 2ie^{S}\exp\left(-\frac{1}{\hbar}\int_a^x \rho(x')dx'\right)\right]. \tag{8.280}$$

2

$$\frac{1}{\hbar}\int_x^b \rho dx = \frac{1}{\hbar}\int_a^b \rho dx - \frac{1}{\hbar}\int_a^x \rho dx.$$

.

8.6.4.1 Connection at a

Now we connect this wave function to the left of the point $x = a$. Using the connection formulae:

$$\frac{1}{\sqrt{\rho}} \exp\left(-\frac{1}{\hbar}\int_a^x \rho(x')dx'\right) \to \frac{2}{\sqrt{p}} \sin\left[\frac{1}{\hbar}\int_x^a p(x')dx' + \frac{\pi}{4}\right] \tag{8.281}$$

and

$$\frac{1}{\sqrt{\rho}} \exp\left(\frac{1}{\hbar}\int_a^x \rho(x')dx'\right) \to \frac{1}{\sqrt{p}} \cos\left[\frac{1}{\hbar}\int_x^a p(x')dx' + \frac{\pi}{4}\right], \tag{8.282}$$

we have the wave function at $x < a$

$$\phi(x < a) = \frac{4C}{i\sqrt{p}} e^S \sin\left[\frac{1}{\hbar}\int_x^a p(x')dx' + \frac{\pi}{4}\right] + \frac{C}{\sqrt{p}} e^{-S} \cos\left[\frac{1}{\hbar}\int_x^a p(x')dx' + \frac{\pi}{4}\right].$$

$$= \frac{4C}{i\sqrt{p}} e^S \sin\left[-\frac{1}{\hbar}\int_a^x p(x')dx' + \frac{\pi}{4}\right] + \frac{C}{\sqrt{p}} e^{-S} \cos\left[-\frac{1}{\hbar}\int_a^x p(x')dx' + \frac{\pi}{4}\right]. \tag{8.283}$$

Using the notation (8.272), we have

$$\phi(x < a) = \frac{4C}{i\sqrt{p}} e^S \sin[-Z] + \frac{C}{\sqrt{p}} e^{-S} \cos[-Z]. \tag{8.284}$$

This is rewritten as[3]

$$= \frac{2C}{\sqrt{p}} \left(\left(e^S + \frac{e^{-S}}{4}\right) e^{iZ} - \left(e^S - \frac{e^{-S}}{4}\right) e^{-iZ}\right). \tag{8.285}$$

Thus, the coefficients A and A' are[4]

3

$$-\frac{4}{i} e^S \frac{e^{iZ} - e^{-iZ}}{2i} + e^{-S} \frac{e^{iZ} + e^{-iZ}}{2} = 2e^S (e^{iZ} - e^{-iZ}) + e^{-S} \frac{e^{iZ} + e^{-iZ}}{2}$$
$$= 2\left[e^{iZ}(e^S + \frac{1}{4}e^{-S}) - e^{-iZ}(e^S - \frac{1}{4}e^{-S})\right].$$

4 It should be noted that the wave functions in the present argument are given in the form

$$e^{\frac{1}{\hbar}\int_a^x p(x')dx' - \frac{\pi}{4}},$$

but not

$$e^{\frac{1}{\hbar}\int_a^x p(x')dx'}.$$

$$A = e^S + \frac{e^{-S}}{4}, \quad A' = -\left(e^S - \frac{e^{-S}}{4}\right). \tag{8.286}$$

The transmission coefficient is given by

$$T = \left| \frac{2C}{2C\left(e^S + \frac{e^{-S}}{4}\right)} \right|^2 = e^{-2S}\left(1 + \frac{e^{-2S}}{4}\right)^{-2}. \tag{8.287}$$

When S is large, $e^{-2S} \ll 1$ and the transmission coefficient is given by

$$T \simeq \exp\left(-\frac{2}{\hbar}\int_a^b \sqrt{2m(V(x') - E)}dx'\right). \tag{8.288}$$

Initial relaxation of the metastable state is given by this transmission.

8.6.4.2 The Other Wave Function

Here, we also obtained the independent wave function. By (8.273), the independent wave function of the form

$$\phi_{\text{out}}(x) = \frac{2C}{\sqrt{p(x)}} \exp\left[-\frac{i}{\hbar}\int_b^x p(x')dx' + i\frac{\pi}{4}\right], \quad x > b \tag{8.289}$$

is connected to

$$\phi_{\text{barrier}}(x) = \frac{C}{\sqrt{\rho(x)}}\left[\exp\left(-\frac{1}{\hbar}\int_x^b \rho(x')dx'\right) + 2i\exp\left(\frac{1}{\hbar}\int_x^b \rho(x')dx'\right)\right]. \tag{8.290}$$

Using (8.279), we have

$$\phi_{\text{barrier}}(x) = \frac{C}{\sqrt{\rho(x)}}\left[e^{-S}\exp\left(\frac{1}{\hbar}\int_a^x \rho(x')dx'\right) + 2ie^S\exp\left(-\frac{1}{\hbar}\int_a^x \rho(x')dx'\right)\right]. \tag{8.291}$$

Using the same connection formulae: we have the wave function at $x < a$

$$\phi(x<a) = -\frac{4C}{i\sqrt{p}}e^{S}\sin\left[\frac{1}{\hbar}\int_x^a p(x')dx' + \frac{\pi}{4}\right] + \frac{C}{\sqrt{p}}e^{-S}\cos\left[\frac{1}{\hbar}\int_x^a p(x')dx' + \frac{\pi}{4}\right].$$

$$= -\frac{4C}{i\sqrt{p}}e^{S}\sin\left[-\frac{1}{\hbar}\int_a^x p(x')dx' + \frac{\pi}{4}\right] + \frac{C}{\sqrt{p}}e^{-S}\cos\left[-\frac{1}{\hbar}\int_a^x p(x')dx' + \frac{\pi}{4}\right].$$

$$= \frac{2C}{\sqrt{p}}\left(\left(e^{S} + \frac{e^{-S}}{4}\right)e^{-iZ} - \left(e^{S} - \frac{e^{-S}}{4}\right)e^{iZ}\right). \tag{8.292}$$

8.6.5 Transition Matrix

Here, we consider a general connection formula over a barrier depicted in Fig. 8.7.

The wave function

$$\Psi(x>b) = \frac{2C}{\sqrt{p(x)}}\exp\left[\frac{i}{\hbar}\int_b^x p(x')dx' - i\frac{\pi}{4}\right] + \frac{2C'}{\sqrt{p(x)}}\exp\left[-\frac{i}{\hbar}\int_b^x p(x')dx' + i\frac{\pi}{4}\right] \tag{8.293}$$

is connected to

$$\Psi(x<a) = \frac{2C}{\sqrt{p}}\left(\left(e^{S} + \frac{e^{-S}}{4}\right)e^{iZ} - \left(e^{S} - \frac{e^{-S}}{4}\right)e^{-iZ}\right)$$

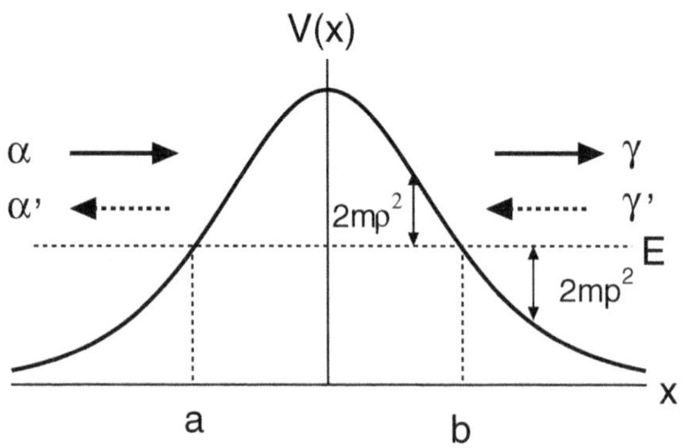

Fig. 8.7 Tunneling through an energy barrier $V(x)$. Here, $\psi(x>b) = \frac{\gamma}{\sqrt{p(x)}}\exp\left[\frac{i}{\hbar}\int_b^x p(x')dx'\right] + \frac{\gamma'}{\sqrt{p(x)}}\exp\left[-\frac{i}{\hbar}\int_b^x p(x')dx'\right]$ and $\phi(x<a) = \frac{\alpha}{\sqrt{p}}\exp\left[\frac{i}{\hbar}\int_a^x p(x')dx'\right] + \frac{\alpha'}{\sqrt{p}}\exp\left[-\frac{i}{\hbar}\int_a^x p(x')dx'\right]$

$$+\frac{2C'}{\sqrt{p}}\left(\left(e^S+\frac{e^{-S}}{4}\right)e^{-iZ}-\left(e^S-\frac{e^{-S}}{4}\right)e^{iZ}\right). \tag{8.294}$$

Expressing this as

$$\Psi(x<a)=\frac{2}{\sqrt{p}}\left(Ae^{iZ}+A'e^{-iZ}\right), \tag{8.295}$$

the coefficients are

$$\frac{2}{\sqrt{p}}A=\frac{2C}{\sqrt{p}}\left(e^S+\frac{e^{-S}}{4}\right)-\frac{2C'}{\sqrt{p}}\left(e^S-\frac{e^{-S}}{4}\right), \tag{8.296}$$

$$\frac{2}{\sqrt{p}}A'=-\frac{2C}{\sqrt{p}}\left(e^S-\frac{e^{-S}}{4}\right)+\frac{2C'}{\sqrt{p}}\left(e^S+\frac{e^{-S}}{4}\right). \tag{8.297}$$

The relation of coefficients is expressed in the matrix representation:

$$\begin{pmatrix}A\\A'\end{pmatrix}=M\begin{pmatrix}C\\C'\end{pmatrix}, \tag{8.298}$$

where

$$M=\begin{pmatrix}\left(e^S+\frac{e^{-S}}{4}\right) & -\left(e^S-\frac{e^{-S}}{4}\right)\\ -\left(e^S-\frac{e^{-S}}{4}\right) & \left(e^S+\frac{e^{-S}}{4}\right)\end{pmatrix}. \tag{8.299}$$

The inverse connection is given by

$$\begin{pmatrix}C\\C'\end{pmatrix}=M^{-1}\begin{pmatrix}A\\A'\end{pmatrix}, \tag{8.300}$$

where[5]

$$M^{-1}=\begin{pmatrix}\left(e^S+\frac{e^{-S}}{4}\right) & -\left(e^S-\frac{e^{-S}}{4}\right)\\ -\left(e^S-\frac{e^{-S}}{4}\right) & \left(e^S+\frac{e^{-S}}{4}\right)\end{pmatrix}. \tag{8.301}$$

If we express (8.293) as

5

$$\begin{pmatrix}a&b\\b&a\end{pmatrix}\begin{pmatrix}a&-b\\-b&a\end{pmatrix}=(a^2-b^2)\begin{pmatrix}1&0\\0&1\end{pmatrix}.$$

$$a=e^S+\frac{e^{-S}}{4},\quad b=-\left(e^S-\frac{e^{-S}}{4}\right),\quad a^2-b^2=1.$$

.

$$\Psi(x>b)=\frac{2\gamma}{\sqrt{p(x)}}\exp\left[\frac{i}{\hbar}\int_b^x p(x')dx'\right]+\frac{2\gamma'}{\sqrt{p(x)}}\exp\left[-\frac{i}{\hbar}\int_b^x p(x')dx'\right], \tag{8.302}$$

the coefficients are related as

$$\begin{pmatrix}\gamma\\ \gamma'\end{pmatrix}=\begin{pmatrix}e^{-i\frac{\pi}{4}} & 0\\ 0 & e^{i\frac{\pi}{4}}\end{pmatrix}\cdot\begin{pmatrix}C\\ C'\end{pmatrix}. \tag{8.303}$$

And if we express (8.295) as

$$\Psi(x<a)=\frac{2}{i\sqrt{p}}\left(\alpha\exp\left[\frac{i}{\hbar}\int_b^x p(x')dx'\right]+\alpha'\exp\left[-\frac{i}{\hbar}\int_b^x p(x')dx'\right]\right), \tag{8.304}$$

the coefficients are related as

$$\begin{pmatrix}\alpha\\ \alpha'\end{pmatrix}=\begin{pmatrix}e^{-i\frac{\pi}{4}} & 0\\ 0 & e^{i\frac{\pi}{4}}\end{pmatrix}\begin{pmatrix}A\\ A'\end{pmatrix}. \tag{8.305}$$

Thus, the relation (8.306) is

$$\begin{pmatrix}e^{i\frac{\pi}{4}} & 0\\ 0 & e^{-i\frac{\pi}{4}}\end{pmatrix}\begin{pmatrix}\alpha\\ \alpha'\end{pmatrix}=M\begin{pmatrix}e^{i\frac{\pi}{4}} & 0\\ 0 & e^{-i\frac{\pi}{4}}\end{pmatrix}\begin{pmatrix}\gamma\\ \gamma'\end{pmatrix}. \tag{8.306}$$

Therefore, the transition matrix of coefficients from (8.302) to (8.304) is

$$M'=\begin{pmatrix}e^{-i\frac{\pi}{4}} & 0\\ 0 & e^{i\frac{\pi}{4}}\end{pmatrix}M\begin{pmatrix}e^{i\frac{\pi}{4}} & 0\\ 0 & e^{-i\frac{\pi}{4}}\end{pmatrix}.$$

$$=\begin{pmatrix}\left(e^S+\frac{e^{-S}}{4}\right) & i\left(e^S-\frac{e^{-S}}{4}\right)\\ -i\left(e^S-\frac{e^{-S}}{4}\right) & \left(e^S+\frac{e^{-S}}{4}\right)\end{pmatrix}. \tag{8.307}$$

$$(M')^{-1}=\begin{pmatrix}\left(e^S+\frac{e^{-S}}{4}\right) & -i\left(e^S-\frac{e^{-S}}{4}\right)\\ i\left(e^S-\frac{e^{-S}}{4}\right) & \left(e^S+\frac{e^{-S}}{4}\right)\end{pmatrix}. \tag{8.308}$$

8.6.6 Relaxation from Metastable Potential

Next, we consider the potential depicted in Fig. 6.12. Using the connection formula from (8.267) to (8.268), it is connected to

$$\psi_1(x>c)=\frac{2D}{\sqrt{p(x)}}\cos\left[\frac{1}{\hbar}\int_c^x p(x')dx'-\frac{\pi}{4}\right]$$

$$=\frac{1}{\sqrt{p(x)}}\left(D\exp\left(\frac{i}{\hbar}\int_c^x p(x')dx\right)e^{-i\frac{\pi}{4}}+D\exp\left(-\frac{i}{\hbar}\int_c^x p(x')dx\right)e^{i\frac{\pi}{4}}\right). \tag{8.309}$$

Now we consider the wave function using the integral between x and a.

$$\exp\left(\frac{i}{\hbar}\int_c^x p(x')dx\right)=\exp\left(\frac{i}{\hbar}\int_c^a p(x')dx-\frac{i}{\hbar}\int_x^a p(x')dx\right)$$

$$=\exp\left(\frac{i}{\hbar}\int_c^a p(x')dx+\frac{i}{\hbar}\int_a^x p(x')dx\right). \tag{8.310}$$

Setting

$$X\equiv\frac{1}{\hbar}\int_c^a p(x')dx', \tag{8.311}$$

$\psi_1(x>c)$ is rewritten as

$$\psi_1(x>c)=\frac{D}{\sqrt{p(x)}}\left(\exp\left(\frac{i}{\hbar}\int_a^x p(x')dx\right)e^{-i\frac{\pi}{4}}e^{iX}\right.$$
$$\left.+\exp\left(-\frac{i}{\hbar}\int_a^x p(x')dx\right)e^{i\frac{\pi}{4}}e^{-iX}\right). \tag{8.312}$$

Using the connection formula over the barrier $(M')^{-1}$ (8.308),

$$\begin{pmatrix}\gamma\\ \gamma'\end{pmatrix}=\begin{pmatrix}\left(e^S+\frac{e^{-S}}{4}\right) & -i\left(e^S-\frac{e^{-S}}{4}\right)\\ i\left(e^S-\frac{e^{-S}}{4}\right) & \left(e^S+\frac{e^{-S}}{4}\right)\end{pmatrix}\begin{pmatrix}\alpha\\ \alpha'\end{pmatrix}, \tag{8.313}$$

where

$$\alpha = e^{-i\frac{\pi}{4}} e^{iX}, \quad \alpha' = e^{i\frac{\pi}{4}} e^{-iX}. \tag{8.314}$$

Explicitly calculating, we have the relation

$$\begin{pmatrix} \gamma \\ \gamma' \end{pmatrix} = \begin{pmatrix} \left(e^S + \frac{e^{-S}}{4}\right) e^{-i\frac{\pi}{4}} e^{iX} - i \left(e^S - \frac{e^{-S}}{4}\right) e^{i\frac{\pi}{4}} e^{-iX} \\ i \left(e^S - \frac{e^{-S}}{4}\right) e^{-i\frac{\pi}{4}} e^{iX} + \left(e^S + \frac{e^{-S}}{4}\right) e^{i\frac{\pi}{4}} e^{-iX} \end{pmatrix}. \tag{8.315}$$

References

1. Kramers, H. (1940). *Physica Amsterdam, 7*, 284.
2. Langer, J. S. (1967). *Annals of Physics, 41*, 108–157.
3. Langer, J. S. (1969). *Annals of Physics, 54*, 258–275.
4. Callen, C., & Coleman, S. (1977). *Physical Review D, 16*, 1762; Coleman, S. (1885). *Aspects of symmetry*. Cambridge University Press.
5. Affleck, Ian. (1981). Quantum-statistical metastability. *Physical Review Letters, 46*(6), 388–391. https://doi.org/10.1103/PhysRevLett.46.388
6. Owerre, S. A., & Paranjape, M. B. (2015). *Physics Reports, 546*, 1.
7. Ankerhold, J. (2006). *Quntum tunneling in complex systems: The semiclassical approach*. Springer.
8. Lenz, W. (1920). *Physikalische Zeitschrift, 61*, 613; Ising, E. (1925). *Zeitschrift für Physik, 31*, 253.
9. Stoner, E. C., & Wohlfarth, E. P. (1948). *Philosophical Transactions of the Royal Society A, 240*, 599.
10. Bonet, E., Wernsdorfer, W., Barbara, B., Benoit, A., Mailly, D., & Thiaville, A. (1999). *Physical Review Letters, 83*, 4188.
11. Tamura, R., Tanaka, S., & Kawashima, N. (2010). *Progress of Theoretical Physics, 124*, 381–388.
12. Blume, M. (1966). *Physical Review,141*, 517; Capel, H. (1966). *Physica,32*, 966.
13. Kahn, O. (2013). Molecular magnetism (VCII, 1993). In Halcrow, M. A. (ed.), *Spin-crossover materials*. Wiley; Guetlich, G., & Goodwin, H. E., (eds.) (2004). *Spin crossover in transition metal compounds I, II, and III. Topics in current chemistry*, Vols. 233, 234, 235. Springer.
14. Wajnflasz, J. (1970). Physica Status Solidi, 40, 537–545; Wajnflasz. J., & Pick, R. (1971). Journal de Physique Colloques, 32, C1 91–92; Bousseksou, A., Nasser, J., Linares, J., Boukheddaden, K., & Varret, F. (1992). *Journal of Physics I France, 2*, 1381–1403.
15. Miyashita, S., Konishi, Y., Tokoro, H., Nishino, M., Boukheddaden, K., & Varret, F. (2005). *Progress of Theoretical Physics, 114*, 719.
16. Decurtins, S., et al. (1985). *Inorganic Chemistry, 24*, 2174.
17. Konishi, Y., Tokoro, H., Nishino, M., & Miyashita, S. (2006). *Journal of the Physical Society of Japan, 75*, 114603.
18. Tokoro, H., Miyashita, S., Hashimoto, K., & Ohkoshi, S. (2006). *Physical Review B, 73*, 172415.
19. Tokoro, H., & Ohkoshi, S. (2008). *Applied Physics Letters, 93*, 21906.
20. Nishino, M., Boukheddaden, K., Konishi, Y.,& Miyashita, S. (2007) *Physical Review Letters, 98*, 247203. Enachescu, C., Nishino, N., & Miyashita, S. (2013). In Halcrow, M. A. (ed.), *Spin-crosover material* (pp. 455–474). Wiley.

S. Miyashita, *Collapse of Meta stability*, Fundamental Theories of Physics 211, https://doi.org/10.1007/978-981-19-6668-2

21. Miyashita, S., Konishi, Y., Nishino, M., Tokoro, H., & Rikvold, P. A. (2008). *Physical Review B, 77*, 0144105.
22. Nishino, M., Enachescu, C., Miyashita, S., Rikvold, P. A., Boukheddaden, K., & Varret, F. (2011). *Scientific Reports, 1162*, 1–5.
23. Mori, T., Miyashita, S., & Rikvold, P. A. (2010). *Physical Review E, 81*, 011135.
24. Nishino, M., & Miyashita, S. (2013). *Physical Review B, 88*, 014108; Nishino, M., Rikvold, P. A., Omand, C., & Miyashita, S. (2018). *Physical Review B, 98*, 144402.
25. (2015). *Handbook of chemistry and physics* (96th ed.) CRC Press.
26. For example, see the textbook: Reichl, L. E. (1980). *A modern course of statistical physics*. University of Texas Press.
27. Mayer, J. E., & Mayer, M. G. (1941). *Statistical mechnics*. Wiley.
28. Ozeki, Y., Kasono, K., Ito, N., & Miyashita, S. (2003). *Physica A, 321*, 271.
29. Glauber, R. J. (1963). *Journal of Mathematical Physics, 4*, 294.
30. Child, M. S. (2014). *Semiclassical mechanics with molecular applications* (2nd ed.). Oxford University Press.
31. Kubo, R., Matsuo, K., & Kitahara, K. (1973). *Journal of Statistical Physics, 9*, 51; Paul, W., Heermann, D. W., & Binder, K. (1989). *Journal of Physics A, 22*, 3325.
32. Tomita, H., Ito, A., & Kidachi, H. (1976). *Progress of Theoretical Physics, 56*, 786.
33. Tomita, H., & Miyashita, S. (1992). *Physical Review B, 46*(8886), 8886.
34. Rikvold, P. A., Tomita, H., Miyashita, S., & Sides, S. W. (1994). *Physical Review E, 49*, 5080.
35. Néel, L. (1949). *Annals of Geophysics, 5*, 99; see also the review, Barbara, B. (2019). *Compte Rendus Physique, 20*, 631.
36. Brown, W. F. (1963). *Physical Review, 130*, 1677.
37. Kurkijaervi, J. (1972). *Physical Review B, 6*, 832; Gunther, L., & Barbara, B. (1994). *Physical Review B, 49*, 3926.
38. Wernsdorfer, W., Bonet Orozco, E., Hasselbach, K., Benoit, A., Barbara, B., Demoncy, N., et al. (1997). *Physical Review Letters, 78*, 1791.
39. Yomogita, T., Okamoto, S., Kikuchi, N., Kitakami, O., Sepehri-Amin, H., & Takahashi, Y. K. (2020). *Tadakatsu Ohkubo*. Kazuhiro Hono: Keiko Hioki, and Atsushi Hattori, Acta Mater.
40. Bray, A. J. (2002). *Advances in Physics, 51*, 481–587.
41. Suzuki, M. (1977). *Progress of Theoretical Physics, 56*, 77 and 477.
42. Hohenberg, P. C., & Halperin, B. I. (1977). *Reviews of Modern Physics, 49*, 435.
43. Miyashita, S., & Takano, H. (1985). *Progress of Theoretical Physics, 73*, 1122–1140.
44. Griffiths, R. B. (1969). *Physical Review Letters, 23*, 17.
45. Takano, H., & Miyashita, S. (1989). Relaxation of the spin autocorrelation function in the kinetic Ising model with bond dilution. *Journal of the Physical Society of Japan, 58*, 3871–3874.
46. Takano, H., Nakanishi, H., & Miyashita, S. (1988). *Physical Review B, 37*, 3716–3719.
47. Huse, D. A., & Fisher, D. S. (1987). *Physical Review B, 35*, 6841.
48. Tang, C., Nakanishi, H., & Langer, J. S. (1989). *Physical Review A, 40*, 995–1003.
49. Lifshitz, I. M., & Slyozov, V. V. (1961). Journal of Physics and Chemistry of Solids, 19, 35–50; Wagner, C. (1961). *Zeitschrift für Elektrochemie, 65*, 581–591.
50. Jinnai, H., Nishikawa, Y., Morimoto, H., Koga, T., & Hashimoto, T. (2000). *Langmuir, 16*(9), 4380–4393.
51. Gromadzki, D. (2010). *Reviews in Environmental Science and Biotechnology, 9*, 301–306.
52. Ostwald, W. (1896). Lehrbuch der Allgemeinen Chemie (Vol. 2, part 1). Leipzig, Germany; Ostwald, W. (1897). *Zeitschrift für physikalische Chemie, 22*, 289–330.
53. Baldan, A. (2002). *Journal of Materials Science, 37*, 2171–2202.
54. Schultz, T. D., Mattis, D. C., & Lieb, E. H. (1964). *Reviews of Modern Physics, 36*, 846.
55. Gel'fant, I. M., & Yaglom, A. M. (1960). *Journal of Mathematical Physics, 1*, 48.
56. Brush, S. G. (1961). *Reviews of Modern Physics, 33*, 79–92.
57. Kac, M. (1949). *Transactions of the American Mathematical Society, 65*, 1.
58. Burton, W. K., & de Rorde, A. H. (1955). *Nouvo cimento, 2*, 197.

59. Hatano, N., Sasada, K., Nakamura, H., & Petrosky, T. (2008). *Progress of Theoretical Physics, 119*, 187.
60. Landau, L. D., & Lifshitz, E. M. (1965). *Quantum mechanics*. Pergamon Press.
61. Stamp, P. C. E., Chudonovsky, E. M., & Barbara, B. (1992). Quantum tunneling of magnetization in solid. *International Journal of Modern Physics, 6*.
62. Barbara, B., Sampaio, L. C., Wegrowe, J. E., Ratnam, B. A., Marchand, C., Paulsen, A. et al. (1993). *Journal of Applied Physics, 73* 6703; Egami, T. (1973). *Physica Status Solidi, 29*, 157; *57*, 211.
63. Barbara, B., & Wernsforfer, W. (1997). *Current Opinian in Solid State & Material Science, 2* 220; Barbara, B., Paulsen, C., Sampaio, L. C., Uehara, M., Fruchard, F., Tholence, J.-L., & Marchand, A. (1992). In Dormann, J. L. (Ed.), *Proceedings of International Workshop Studies of Magnetic Properties of Fine Particles, 1991 Rome* (pp. 235–242). Elsevier.
64. Kato, Y., & Tanaka, A. (1994). *Journal of the Physical Society of Japan, 63*, 1277.
65. Miyashita, S. (2007). *Journal of the Physical Society of Japan, 76*, 104003.
66. Siegert, A. J. F. (1939). *Physical Review, 56*, 750.
67. Feshbach, H. (1958). *Annals of Physics, 5*, 357; (1962). *Annals of Physics, 19*, 287.
68. Fano, U. (1961). *Physical Review, 124*, 1866.
69. Hazi, A. U., & Taylor, H. S. (1970). *Physical Review A, 1*, 1109.
70. Langer, R. E. (1937). *Physical Review, 51*, 669.
71. Miller, S. C., & Good, R. H. (1953). *Physical Review, 91*, 174.
72. Connor, J. N. L. (1968). *Molecular Physics, 15*, 621.
73. Dickinson, A. S. (1970). *Molecular Physics, 18*, 441.
74. Meyer, H.-D., & Walter, O. (1982). *Journal of Physics B: Atomic and Molecular Physics, 15*, 3647.
75. Leggett, A. J., Chakravarty, S., Dorsey, A. T., Fisher, M. P. A., Garg, A., & Zwerger, W. (1987). *Reviews of Modern Physics, 59*, 1.
76. Caldeira, A. O., & Leggett, A. J. (1981). *Physical Review Letters, 46*, 211 (1981); Caldeira, A. O., & Leggett, A. J. (1985). *Annals of Physics, 149*, (1983) 374; Caldeira, A. O., & Leggett, A. J. (1985). *Physical Review A, 31*, 1059–1066.
77. Morita, S., Teranishi, Y., Hatano, N., & Miyashita, S., in preparation.
78. Teranishi, private communication.
79. Morita, S. (2007). *Journal of the Physical Society of Japan, 76*, 104001.
80. Stevens, K. W. H. (1952). *Proceedings of the Physical Society A, 65*, 299 (1952); Görrler-Walrand, C. & Binnemans, K. (1996). Rationalization of crystal-field parametrization. In Gschneidner, K. A., & Eyring, L. (eds.), *Handbook on the physics and chemistry of rare earths* (Vol. 23). North Holland Publishing Company.
81. Wernsdorfer, W., Bonet Orozco, E., Hasselbach, K., Benoit, A., Barbara, B., Demoncy, N., et al. (1977). *Physical Review Letters, 78*, 1791; Wernsdorfer, W., Hasselbach, K., Mailly, D., Barbara, B., Benoit, A., Thomas, L., & Suran, G. (1995). *Journal of Magnetism and Magnetic Materials,145*, 33.
82. Thomas, L., Lionti, F., Ballou, R., Gatteschi, D., Sessoli, R., & Barbara, B. (1996). Nature *383*, 145; Perenboom, J. A. A. (1998). *Physical Review B, 58*, 330; Gatteschi, D., Sessoli, R., & Villain, J. (2006). *Molecular nanomagnets*. Oxford University Press.
83. Barbara, B., Giraud, R., Wernsdorfer, W., Mailly, D., Lejay, P., Tkachuk, A., et al. (2004). *Journal of Magnetism and Magnetic Materials, 272–276*, 1024.
84. Landau, L. D. (1932). *Physikalische zeitschrift der Sowjetunion, 2*, 46.
85. Zener, C. (1932). *Proceedings of the Royal Society of London. Series A, 137*, 696.
86. Stuekelberg, E. C. G. (1932). *Helvetica Physica Acta, 5*, 369. Majorana, E. (1932). *Nuovo Cimento, 9*, 43.
87. Miyashita, S. (1995). *Journal of the Physical Society of Japan, 64*, 3214.
88. Miyashita, S. (1996). *Journal of the Physical Society of Japan, 65*, 2734.
89. Wernsdorfer, W., & Sessoli, R. (1999). *Science, 284*, 133.
90. Ueda, M., Maegawa, S., & Kitagawa, S. (2002). *Physical Review B, 66*, 073309.
91. Saito, K., Miyashita, S., & De Raedt, H. (1999). *Physical Review B, 60*, 14553.

92. Saito, K., & Miyashita, S. (2001). *Journal of the Physical Society of Japan, 70*, 3385.
93. Shapira, Y., Liu, M. T., Foner, S., Dubé, C. E., & Bonitatebus, P. J. Jr. (1999). *Physical Review B, 59*, 1046.
94. Nakano, H., & Miyashita, S. (2001). *Journal of the Physical Society of Japan, 70*, 2151.
95. Chiorescu, I., Wernsdorfer, W., Müller, A., Bögge, H., & Barbara, B. (2000). *Physical Review Letters, 84*, 3454.
96. Kayanuma, Y. (1982). *Journal of the Physical Society of Japan, 51*, 3526.
97. Kayanuma, Y. (1984). *Journal of the Physical Society of Japan, 53*, 108.
98. Phys, J. (1985). *Society of Japan, 54*, 2037.
99. Ao, P., & Rammer, J. (1989). *Physical Review Letters, 62*, 3004; (1991). *Physical Review B, 43*, 5397.
100. Kayamuma, Y. (1998). *Physical Review B, 47*, 9940.
101. Kayanuma, Y., & Nakayama, H. (1993). *Physical Review B, 57*, 13099.
102. Kayanuma, Y., & Nakamura, H. (1998). *Physical Review B, 57*, 13099; Hernandez, J. M., Zhang, X. X., Luis, F., Tejada, T., FriedGman, J. R., Sarachik, M. P., et al. (1997). *Physical Review B, 56*, 11102.
103. Saotoh, K., & Kayanuma, Y. (1997). *Physical Review A, 65*, 033407.
104. Nishino, M., Onishi, H., Yamaguchi, K., & Miyashita, S. (2000). *Physical Review B, 62*, 9463–9471.
105. Dodin, A., Garmon, S., Simine, L., & Segal, D. (2014). *Journal of Chemical Physics, 140*, 124709.
106. Feyman, R. P., & Vernon, F. L. (1963). *Annals of Physics, 24*, 118.
107. *Physical Review Letters, 49*, 1545 (1982).
108. van Hemmen, J. L., & Suto, A. (1995). In Gunter, L., & Barbara, B. (eds.), *Quantum tunneling of magnetism QTM94*. NATO Advanced Institutes, Ser. E. (Vol. 301, p. 19). Kluwer, Dordrecht.
109. Florens, S., Venturelli, D., & Narayanan, R. (2010). In Chandra, A. K., et al. (eds.), *Quantum quenching, annealing and computation* (pp. 145–162). Springer.
110. V ölker, K. (1998). *Physical Review B, 58*, 1862.
111. Strathearn, A., Kirton, P., Kilda, D., Keeping, L., & Lovett, B. W. *Nature Communication*. s41467-018-05617-3.
112. Saito, K., & Kato, T. (2013). *Physical Review Letters, 111*, 214301.
113. Hatomura, T., Barbara, B., & Miyashita, S. (2016). *Physical Review Letters, 116*, 037203.
114. Garanin, D. A., & Schilling, R. (2004). *Physical Review B, 69*, 104412.
115. Hatomura, T., Barbara, B., & Miyashita, S. (2017). *Physical Review B, 96*, 134009.
116. Miyashita, S., De Raedt, H., & Barbara, B. (2009). *Physical Review B, 79*, 104422.
117. Suzuki, S., Inoue, J., & Chakrabarti, B. K., (2012). *Quantum ising phases and transitions in transverse ising models*. Lecture notes in physics (Vol. 862). Springer.
118. Suzuki, M. (1976). *Communications in Mathematical Physics, 51*, 183; (1985). *Physical Review B, 31*, 2957.
119. Kadowaki, T., & Nishimori, H. (1998). *Physical Review E, 58*, 5355; Das, A., & Chakrabarti, B. K. (2008). textitReviews of Modern Physics, *80*, 1061; Santoro, G., & Tosatti, E. (2006). *Journal of Physics A, 39*, R393.
120. D-wave system: https://www.dwavesys.com.
121. Dicke, R. H. (1954). *Physical Review, 93*, 99.
122. Jaynes, E., & Cummings, F. W. (1963). *Proceedings of the IEEE, 51*, 89.
123. Tavis, M., & Cummings, F. W. (1968). *Physical Review, 170*, 379.
124. Walther, H., Varcoe, B. T. H., Englert, B.-G., & Becker, T. (2006). *Reports on Progress in Physics, 69*(5), 1325–1382; Miller, R., Northup, T. E., Birnbaum, K. M., Boca, A., Boozer, A. D., & Kimble, H. J. (2005). *Journal of Physics B: Atomic, Molecular and Optical Physics, 38*(9), S551–S565.
125. Kubo, R., & Tomita, K. (1954). *9*, 888. Kubo, R. (1957). *Journal of the Physical Society of Japan, 12*, 570.
126. Agarwal, G. S. (1984). *Physical Review Letters, 53*, 1732; Houdre, R., Stanley, R. P., & Ilegems, M. (1996). *Physical Review A, 53*, 2711. Zhu, Y., Gauthier, D. J., Morin, S. E., Wu, Q., Carmichael, H. J., & Mossberg, T. W. (1990). *Physical Review Letters, 64*, 2499.

127. Chiorescu, I., Groll, N., Bertaina, S., Mori, T., & Miyashita, S. (2010) Magnetic strong coupling in a spin-photon system and transition to classical regime. *Physical Review B, 82*(1–7), 024413.
128. Miyashita, S., Shirai, T., Mori, T., De Raedt, H., Bertaina, S., & Chiorescu, I. (2012). *Journal of Physics B: Atomic, Molecular and Optical Physics, 45*, 124010(1–10).
129. Kubo, Y., Ong, F. R., Bertet, P., Vion, D., Jacques, V., Zheng, D., et al. (2010). *Physical Review Letters, 105*, 140502; Kubo, Y., Grezes, C., Dewes, A., Umeda, T., Isoya, J., Sumiya, H., Morishita, N., et al. (2011). *Physical Review Letters, 107* 220501. Schuster, D. I., Sears, A. P., Ginossar, E., DiCarlo, L., Frunzio, L., Morton, J.J.L., et al. (2010). *Physical Review Letters, 105*, 140501; Amsüss, R., Koller, Ch., Nöbauer, T., Putz, S., Rotter, S., Sandner, K., et al. (2011). *Physical Review Letters, 107*, 220502; Zhu, X., Saito, S., Kemp, A., Kakuyanagi, K., Karimoto, S.-i., Nakano, H. et al. (2011). *Nature, 478*, 221.
130. Hepp, K., & Lieb, E. H. (1973). *Physical Review A, 8*, 2517.
131. Shirai, T., Mori, T., & Miyashita, S. (2014). Novel symmetry-broken phase in a driven cavity system in the thermodynamic limit. *Journal of Physics B, 47*(1–9), 025501.
132. Rzażewski, K., Wódkiewicz, K. & Żakowicz, W. (1975). *Physical Review Letters, 35*, 432.
133. Gibbs, H. M., McCall, S. L., & Venkatesan, T. N. C. (1976). *Physical Review Letters, 36*, 1135; Felber, F., & Marburger, J. (1976). *Applied Physics Letters, 28*, 731; Lugiato, L. A. (1984). *Progress in Optics, 21*, 69.
134. Rempe, G., Thompson, R. J., Brecha, R. J., Lee, W. D., & Kimble, H. J. (1991). *Physical Review Letters, 67*, 1727; Gripp, J., Mielke, S. L., Orozco, L. A., & Carmichael, H. J. (1996). *Physical Review A, 54*, R3746.
135. Shirai, T., Todo, S., De Raedt, H., & Miyashita, S. (2018). *Physical Review A, 98*, 043802.
136. Mori, T. (2013). *Journal of Statistical Mechanics: Theory and Experiment, 2013*, P06005.
137. Carmichael, H. J. (2008). *Statistical methods in quantum optics 2: Nonclassical fields*. Springer.
138. Sarkar, S., & Satchell, J. S. (1987). *Europhysics Physics Letters, 3*, 797.
139. Lee, T. E., Haffner, H., & Cross, M. C. (2012). *Physical Review Letters, 108*, 023602.
140. Gegg, M., & Richter, M. (2016). *New Journal of Physics, 18*, 043037.
141. Fletcher, R. (1976). In Watson, G. A. (ed.), *Numerical analysis: Proceedings of the dundee conference on numerical analysis* (pp. 73–89). Springer.
142. Shore, B. W., & Knight, P. L. (1993). *Journal of Modern Optics, 40*, 1195.
143. Shirai, T., Todo, S., & Miyashita, S. (2020). *Physical Review A, 101*, 013809.
144. Cirac, J. I. (1992). *Physical Review A, 46*, 4354.
145. Tomé, T., Oliveira, M. J. (1990). *Physical Review A, 41*, 4251; Sides, S. W., Rikvold, P. A., & Novotny, M. A. (1998). *Physical Review Letters, 81*, 834.
146. Sagawa, M., Fujimura, S., Togawa, N., Yamamoto, H., & Matsuura, Y. (1984). *Journal of Applied Physics, 55*, 2083. https://doi.org/10.1063/1.333572; Sagawa, M., & Hirosawa, S. (1988). *Journal of Materials Research, 3*, 45
147. Hirosawa, S., & Hono, K. (2022). Foreword to the focus issue: Science and technology of element-strategic permanent magnets. *Science and Technology of Advanced Materials, 23*, 64–65. https://doi.org/10.1080/14686996.2022.2027641; Coey, J. M. D. (2020). Perspective and prospects for rare earth permanent magnets. *Engineering, 6*, 119–131. https://doi.org/10.1016/j.eng.2018.11.034; Hirosawa, S. (2019). Permanent magnets beyond Nd-Dy-Fe-B. *IEEE Transactions on Magnetics, 55*, 2100506, https://doi.org/10.1109/TMAG.2018.2863737.
148. Miyashita, S., Nishino, M., Toga, Y., Hinokihara, T., Uysal, I. E., Miyake, T., et al. (2021). *Science and Technology of Advanced Materials (STAM), 22*, 658.
149. Street, R., & Woolley, J. C. (1949). *Proceedings of the Physical Society, A62*, 562; Street, R., & Woolley, J. C. (1952). *Proceedings of the Physical Society, B65*, 679; Uehara, M., Barbara, B. (1986). *Journal of Physics, 47*, 235.
150. Givord, D., Lienard, A., Tenaud, P., & Viadieu, T. (1987). *Journal of Magnetism and Magnetic Materials, 67*, L281; Givord, D., Lu, Q., Rossignol, M. F., Tenaud, P., & Viadieu, T. (1990). *Journal of Magnetism and Magnetic Materials, 83*, 183.

151. For example, Ren, W., & Vanden-Eijnden, E. (2007). *Journal of Chemical Physics, 126*, 164103 (2007).
152. For example, Dittrich, R., Schrefl, T., Suess, D., Scholz, W., Forster, H., & Fidler, J. (2002). *Journal of Magnetism and Magnetic Materials, 250*, L12.
153. Toga, Y., Miyashita, S., Sakuma, A., & Miyake, T. (2020). *npj Computational Materials, 6*(1–7), 67.
154. Swendsen, R. H., & Wang, J.-S. (1987). *Physical Review Letters, 58*, 86; U. Wolff (1989). *Physical Review Letters, 62*, 361. Ferrenberg, A. M., & Swendsen, R. H. (1988). *Physical Review Letters, 61*, 2635; *63*, 1195 (1989); Berg, B. A., & Neuhaus, T. (1992). *Physical Review Letters, 68*, 9; (1991). *Physics Letters B, 267*, 249; Lee, J. (1993). *Physical Review Letters, 71*, 211; Berg, B. A. (1996). *Journal of Statistical Physics, 82*, 323; (1998). *Nuclear Physics B, 63*, 982.
155. Wang, F., & Landau, D. P. (2001). *Physical Review Letters, 86*, 2050.
156. Gaunt, P. (1986). *Journal of Applied Physics, 59*, 4129–4132.
157. Givord, D., Lienard, A., Tenaud, P., & Viadieu, T., Magn, T. *Journal of Magnetism and Magnetic Materials, 67*, L281–L285; Givord, D., Lu, Q., Rossignol, M. F., Tenaud, P., & Viadieu, T. (1990). *Journal of Magnetism and Magnetic Materials, 83*, 183–188; Okamoto, S., Goto, R., Kikuchi, N., Kitakami, O., Akiya, T., H. Sepehri-Amin, et al. (2015). *Journal of Applied Physics, 118*, 223903; Bance, S., Fischbacher, J., Kovacs, A., Oezeit, H., Reichel, F., & Schrefl, T. (2015). JOM, *67*, 1350–1356.
158. Barbier, J. C. (1954). *Annales de Physique Paris, 9*, 84; Prejean, J. J., & Souletie, J. (1980). *Journal de Phis, 41*, 1335.
159. Wernsdorfer, W., Bonet Orozco, E., Hasselbach, K., Benoit, A., Mailly, D., Kubo, O., et al. (1997). *Physical Review Letters, 79*, 4014; Sappey, R., Vincent, E., Ocio, M., & Hammanm, J. (2000). *Journal of Magnetism and Magnetic Materials, 221*, 87.
160. Wohlfarth, E. P. (1984). *Journal of Physics F, 14*, L155.
161. García-Palacios, J. L., & Lázaro, F. J. (1998). *Physical Review B, 58*, 14937.
162. Nishino, M., & Miyashita, S. (2015). *Physical Review B, 91*, 134411.
163. Nishino, M., Uysal, I. E., Hinokihara, T., & Miyashita, S. (2020). *Physical Review B, 102*, 020413(R).
164. Nishino, M., Uysal, I. E., Hinokihara, T., & Miyashita, S. (2021). *AIP Advances, 11*, 025102.
165. Wernsdorfer, W., Hasselbach, K., Sulpice, A., Benoit, A., Wegrowe, J.-E., Thomas, L. et al. (1996). *Physical Review B, 53*, 3341; Wegrowe, J.-E., Wernsdorfer, W., Thomas, L., Barbara, B., Sulpice, A., Hasselbach, K., et al. (1996). *Physical Review B, 53*, 6536.
166. Hubert, A., & Schaefer, R. (1998). *Magnetic domains*. Springer. Kooy, C., & Enz, U. (1960). *Philips Research Reports, 15*, 7.
167. Hinokihara, T., & Miyashita, S. (2021). *Physical Review B, 103*, 054421.
168. Appel, A. W. (1985). *SIAM Journal on Scientific and Statistical Computing, 6*, 85; Barnes, J., & Hut, P. (1986). *Nature (London), 324*, 446; Carrier, J., Greengard, L., & Rokhlin, V. (1988). *SIAM Journal on Scientific and Statistical Computing, 9*, 669; Makino, J. (1990). *Journal of Computational Physics, 87*, 148. Darden, T., York, D., & Pedersen, L. (1993). *Journal of Chemical Physics, 98*, 10089; Essmann, U., Perera, L., Berkowitz, M. L., Darden, T., Lee, H., & Pedersen, L. G. (1995). *Journal of Chemical Physics, 103*, 8577; Luijten, E., & Blöte, H.W. (1995). *International Journal of Modern Physics C, 6*, 359 ; Hinzke, D., & Nowak, U. (2000). *Journal of Magnetism and Magnetic Materials, 221*, 365.
169. Mak, C. H. (2005). *Journal of Chemical Physics, 122*, 214110.
170. Fukui, K., & Todo, S. (2009). *Journal of Computational Physics, 228*, 2629.
171. Sasaki, Munetaka, & Matsubara, Fumitaka. (2008). Stochastic cutoff method for long-range interacting systems. *Journal of the Physical Society of Japan, 77*(2), 024004. https://doi.org/10.1143/JPSJ.77.024004.
172. Hinokihara, T., Nishino, M., Toga, Y., & Miyashita, S. (2018). *Physical Review B, 97*, 104427.
173. Preisach, F. (1935). *Zeitschrift für Physik, 94*, 277; Della Torre, E. (1999). *Magnetic hysteresis*. IEEE Press.

174. mayergoyz, I. D., Mathematical Models of Hysteresis, *Physical Review Letters,56*, 1518 (1986). Pike, C. R., A.P. Roberts and Verosub, K. L., J Appl. Phys. *85* (1999) 6660. Pike, C. R., Ross, C. A., Scalettar, R. T. and G. Zimany, First-order reversal curve diagram analysis of a perpendicular nickel nanopillar array, *Physical Review B,71*, 134407 (2005).
175. Okamoto, S., Miyazawa, K., Yomogita, T., Kikuchi, N., Kitakami, O., Toyoki, K., et al. (2019). Temperature dependent magnetization reversal process of a Ga-doped Nd-Fe-B sintered magnet based on first-order reversal curve analysis. *Acta Materialia, 178*, 90–98.
176. Matau, F., Nica, V., Postolache, P. P., et al. (2013). Physical study of the Cucuteni pottery technology. *Journal Archaeological Science, 40*, 914–925.
177. Della Torre, E. (1966). Moving Preisach model. *IEEE Transactions on Audio and Electroacoustics, 14*, 86–92.
178. Miyashita, S., in preparation.
179. Slichter, C. P., & Drickamer, H. G. (1972). *Journal of Chemical Physics, 56*, 2142.
180. Feynman, R. P. (1972). *Statistical mechanism*. Benjamin; Feynman, R. P. (1998). *Statistical mechanics*. CRC Press.
181. Peato, D., & Barraco, D. E. (1996). *Revista Mexicana de Fisica, 42*, 145; Falk, H. (1970). *American Journal of Physics, 38*, 858; Callen, H. B. (1985). *Thermodynamics and an introduction to thermostatistics* (2nd ed.). Wiley.
182. Oeksendal, B. K. (2003). *Stochastic differential equations: An introduction with applications*. Springer. Gardiner, C. W. (2004). *Handbook of stochastic methods* (3 ed.). Springer.
183. Whittaker, E. T., & Watson, G. N. (1962). *Mondern analysis*. Cambridge University Press. Abramowitz, M., & Stegun, I. A. (1965). *Handbook of mathematical functions*. Dover.
184. Lindblad, G. (1976). *Commun. Math. Phys., 48*, 119.
185. Breuer, H. P., & Petruccione, F. (2002). *The theory of open quantum systems*. Oxford University Press.
186. Kossakowski, A. (1972). *Reports on Mathematical Physics, 3*(4), 247; Gorini, V., Kossakowski, A., & Sudarshan, E. C. G. (1976). *Journal of Mathematical Physics, 17*(5), 821.
187. Zwanzig, R. (1961). *Physical Review, 124*, 983. Mori, H., & Fujisaka, H. (1973). *Progress of Theoretical Physics, 49*, 764. Mori, H., Fujisaka, H., & Shigematsu, H. (1974). *Progress of Theoretical Physics, 51*, 109.
188. Kubo, R., Toda, M., & Hashitsume, N. (1973). *Statistical physics II*. Springer; Louisell, W. H. (1985). *Quantum statistical properties of radiation*. Wiley.
189. Saito, K., Takesue, S., & Miyashita, S. (2000). *Physical Review E, 61*, 2397.
190. Miyashita, S. (2011). *Journal of Computational and Theoretical Nanoscience, 8*, 919.
191. (1965). *Applied Magnetic Resonance, 1*, 1.
192. Albach, T., et al. (2012). *New Journal of Physics, 14*, 123016.
193. Feyman, R. P. & Hibbs, A. R. (1995). *Quantum mechanics and path integrals*. MacGraw-Hill.
194. Nagaosa, N. (1999). *Quantum field theory in condensed matter physics*. Springer.
195. Onsager, L., & Machlup, S. (1953). *Physical Review, 91*, 1505; Machlup, S., & Onsager, L. *Physical Review, 91*, 1512.
196. Onsager, L. (1931). *Physical Review, 37*, 405; (1931). *38*, 2265.
197. Schiff, L. I. (1968). *Quantum mechanics*. McGrow-Hill Kogakusha.

Index

S. Miyashita, *Collapse of Meta stability*, Fundamental Theories of Physics 211,
https://doi.org/10.1007/978-981-19-6668-2

The manufacturer's authorised representative in the EU is Springer Nature Customer Service Centre GmbH, Europaplatz 3, 69115 Heidelberg, Germany. If you have any concerns regarding our products, please contact ProductSafety@springernature.com

Printed and bound by CPI Group (UK) Ltd, Croydon, CR0 4YY

30/09/2026

02225317-0007